ALSO BY GERALD WEISSMANN

The Woods Hole Cantata

They All Laughed at Christopher Columbus

The Doctor with Two Heads

The Doctor Dilemma

Democracy and DNA

Darwin's Audubon

The Year of the Genome

Times Books

HENRY HOLT AND COMPANY

New York

The
Year of
the
genome

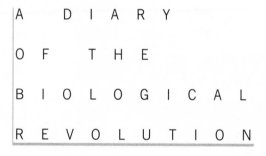

A D I A R Y

O F   T H E

B I O L O G I C A L

R E V O L U T I O N

Gerald Weissmann

Times Books

Henry Holt and Company, LLC

Publishers since 1866

115 West 18th Street

New York, New York 10011

Henry Holt® is a registered trademark
of Henry Holt and Company, LLC.

Frontispiece: Blake, William, "The Ancient of Days," 1794,
relief etching with watercolor, 23.3 x 16.8 cm (9⅛ x 6⅞ in.),
with permission, The British Museum, London

Library of Congress Cataloging-in-Publication Data

Weissmann, Gerald.
  The year of the genome : a diary of the biological revolution /
Gerald Weissmann.—1st ed.
    p. cm.
  Includes index.
  ISBN 0-8050-7095-8 (hc.)
    1. Medical innovations. 2. Medicine—Philosophy. 3. Molecular biology.
  4. Genetics. I. Title.
  R852 .W45 2002
  610—dc21   2001054646

Henry Holt books are available for special promotions and
premiums. For details contact: Director, Special Markets.

First Edition 2002

Designed by Margaret M. Wagner

Printed in the United States of America
10 9 8 7 6 5 4 3 2 1

For Ann, forever

The truth is, that medicine, professedly founded on observation, is as sensitive to outside influences, political, religious, philosophical, imaginative, as is the barometer to the changes of atmospheric density. Theoretically it ought to go on its own straightforward inductive path, without regard to changes of government or to fluctuations of public opinion. But look a moment while I clash a few facts together, and see if some sparks do not reveal by their light a closer relation between the Medical Sciences and the conditions of Society and the general thought of the time, than would at first be suspected.

—OLIVER WENDELL HOLMES, *Medical Essays*

I use quotations like a highway robber, to relieve passersby of their prejudices.　　　　　　　　　　　—WALTER BENJAMIN, *Schriften*

Gerry, let me tell you the difference between business and academia. In business it's dog eat dog. In academia it's the opposite.

—EDWIN C. "JACK" WHITEHEAD

Speed matters.　　　　　　　　　　　　　　—J. CRAIG VENTER

Contents

Contents

The
Year of
the
Genome

Introduction

INTRODUCING the diaries of George Templeton Strong, historian Alan Nevins warned future diarists that "To write a truly great diary one should keep its composition secret and should intend no early publication of its content."[1] This book isn't that sort of diary. My entries provide a personal account of how—in Dr. Holmes's phrase—"medical science interacts with society and the general thought of the time" in the year of the genome. Week by week, over the past year or so, we've been told of another step forward in the progress of science; and week by week we've been hit with news of disease and unreason. These include new outbreaks of Ebola virus, dengue fever, and sub-Saharan AIDS as well as older scourges like typhus, tularemia, and arsenic in our water. The row over RU 486 in America was forgotten amidst worldwide angst over cloning, Dolly, and DDT. Epidemics of mad cow and of foot-and-mouth disease in livestock shared the news with human diseases of memory and desire: Alzheimer's, alcoholism, and drug addiction. Folks "thanked the goddess" for Ralph Nader's showing in the Florida primaries, worried needlessly over a "Balkan syndrome," and hoarded Cipro in fear of anthrax. Slouching in the wings was the rough beast of terror.

On the other hand, it was a year of extraordinarily good news for fans of the Enlightenment. The map of the human genome was substantially completed in record time, an achievement that should rank with the publication of Diderot's *Encyclopédie* or Buffon's *Histoire naturelle*. We've also learned how life on earth began as a flash of light in a cloud of ice; it's not quite Genesis but will do nicely for the secular. The results are in of the first trials of human gene therapy and we've learned that it works. We have crafted a new cure for one form of leukemia, fished out some of the genes that code for long life, and sketched an outline of the chemistry of thought (the 2000 Nobel Prize took note of that). We've been given new ways of handling Parkinson's disease, grinding down the plaques of Alzheimer's, we've turned the tide of AIDS in the developed world and saved hundreds of thousands of lives with a baby aspirin a day. One might say that the biological revolution has opened the way to a medical revolution. These days, we're ready to splice genes in the clinic. Of course, any time a biological scientist feels particularly puffed with pride, along comes an information scientist to prick the bubble. One enthusiast of the new nanotechnology assures us that "If we took your entire genome and burned it onto a CD it would take much less space than Microsoft Office. Lab science is becoming information science—designed on a computer, not at a lab bench. Matter becomes code."[2] So perhaps the nanorevolution is also in the wings.

This then is a diary, or journal, of the year of the genome in which the entries are set in inverse order as they might be set in a photo album. Photo albums mix participants with observers in almost random fashion, the best of diaries use the same method. The photographer becomes the subject, the scenery shifts. We've seen that boy in knickers before, but in khaki; that matron pops up later in bobby socks. There are shots aplenty of the pyramids, St. Paul's, or Machu Picchu, but there's always a familiar figure in the foreground. Let history make the movies, the diary is for snapshots. Caught in the lens of the diarist, on Hilton Head island a Union captain forever salutes Charlotte Forten; in Leamington Spa Miss Loring forever brings tea to

Alice James; at the Century Association, George Templeton Strong grumbles over modern music; in Dresden, Victor Klemperer dodges the Nazis.

With those collections as distant models, I've tried to capture a year in medical science with a wide-angle lens that subtends the general culture. As both participant and observer in the biological revolution, I have no qualms about catching the basic science in necessary detail. People in my field are convinced that while science is only one of the many ways we have of making sense of nature, medical science is the only way we have of making sense of disease. On the other hand, I'm also persuaded that we in our "barbarian moment"—to use Lionel Trilling's phrase—are not the first generation to worry over the stuff of heredity, a plague of vermin, or the dream of eternal life. In that spirit I've called on Ralph Waldo Emerson to explain Craig Venter's drive to unravel our genome, on Voltaire to excuse Rachel Carson's effect on malaria worldwide, and on Aldous Huxley to define the genes of aging at Cal Tech. Clips of poetry and prose by my favorite doctor-writers, O. W. Holmes, Hans Zinsser, and Lewis Thomas, sit cheek by jowl with Auden, Tennyson, and Keith Douglas. Prompted by the newest genes, I trace Patrick O'Brian's heroes to Darwin and Fitzroy; prompted by the newest plagues, I conclude that Anne Frank's death was the logical outcome of Martin Heidegger's belief. That jumble of association is in line with another of my models, the diary/journals kept by Walter Benjamin in which

"it was easy to find next to an obscure love poem from the eighteen century *Der Erste Schnee* a report from Vienna dated 1939 reporting that the local gas company had 'stopped supplying gas to Jews. The gas consumption of the Jewish population involved a loss for the gas company since the biggest customers were the ones who did not pay their bills. The Jews used the gas especially for committing suicide.' "[3]

Happily, the *images noir* of Benjamin and Klemperer are not the theme of this book at all. Over the last year we've seen an efflorescence of biological sciences in the developed world and my overall response is that of guarded triumphalism. Each of those hard-won advances in lab or in clinic was the result of earlier discoveries made in the course of the biological revolution of the last half century, and in each entry of this diary I've attempted to trace the flower to its seed.

In 1976, on the occasion of the First International Congress of Cell Biology held in Boston to celebrate the bicentennial, a group of biologists took note of the occasion to coin the term, "The Biological Revolution." We were eager to distinguish our enterprise from others that had preceded it, such as the sanitarian revolution, the bacteriological revolution, etc. We were proud that thanks to structural biology, electron microscopy, and somatic cell genetics we'd gotten some insights into how cells and tissues are constituted. We also knew that molecular biology—the science of genes and their expression in cells—was at the heart of that revolution. Senator Kennedy, who opened the meeting, caught the temper of the time when he dismissed the naysayers of the Cambridge City Council who had launched "the controversy over recombinant DNA experimentation. Such manipulation of genes is certainly one of the prime examples of the biological revolution at work."[4] The naysayers are still at work, and now they're troubled by stem cells. But while we knew twenty-five years ago that we'd eventually be able to put some of that gene manipulation to work in the clinic, Lewis Thomas rightly confessed that "We [in medicine] have nothing yet to match the astonishing surge of new knowledge achieved by the biological revolution of the past twenty-five years."[5] He would have been pleased to note that at long last the products of that revolution, of that increase in pure knowledge, are beginning to affect the bills of mortality. The Centers for Disease Control and Prevention recently proclaimed that "Deaths from heart disease, cancer, and AIDS declined in 1999." The Centers also noted that deaths from stroke were dramatically reduced.[6] Thus, for the first time since the sanitarian and bacteriological revolutions of 1850–1950, the declines were not due to good plumbing and/or antibiotics but to newer treatments forged in the workshops of the new biology: monoclonal antibodies, antihypertensives, antiretroviral agents—and that baby aspirin every day.

Now that we have all those Gs and Cs and Ts and As spelled out, we've learned from the study of many human genomes that by and large we truly are brothers under the skin. Continents east and continents west, we all came out of Africa together. But a good part of the world is under siege from diseases with which modern medicine alone cannot cope. The forty million or so Africans who are suffering from AIDS are not likely soon to benefit from new chemotherapy, no matter how cheap the treatment. The problem is only in part due to tribal warfare, to the oppression of women, and lack of trained health professionals. I'm convinced, and in this diary I present some of the evidence, that a sanitarian revolution is a necessary preamble to a biological revolution. In the century of the sanitarian revolution, before antibiotics, the only measures Western medicine could deploy against deadly microbes were hygiene and drainage, education and quarantine: soap and sewers, pamphlets and police. Those measures worked. As a result, even before we had diagnostic microbiology, effective vaccines, or antimicrobials, the rates plummeted of cholera, scarlet fever, infantile diarrhea, venereal disease, and tuberculosis. Doctors agreed with Dr. Oliver Wendell Holmes that "the bills of mortality were affected more obviously by drainage than by this or that method of practice." He wrote the anthem of public health for a meeting of the National Sanitary Society in 1860:

> God lent his creatures light and air
> > And waters open to the skies
> Man locks him in a stifling lair
> > And wonders why his brother dies

> In vain our pitying tears are shed
> > In vain we rear our sheltering pile
> Where Art weeds out from bed to bed
> > The plagues we planted by the mile

Be that the glory of the past
> With these our sacred toils begin
So flies in tatters from its mast
> The yellow flag of sloth and sin

And lo! the starry folds reveal
> The blazoned truth we hold so dear
To guard is better than to heal
> The shield is nobler than the spear.[7]

Sad to say, I'd guess that if every medication for HIV and its seque-
lae were available without cost, if every rural village or urban slum
dispensed the pills twice daily, the biggest threat to many patients
would be the water they'd gulp with their pill. The World Health
Organization reports that "In Africa today over half of the population is
without access to safe drinking water and two-thirds lack a sanitary
means of excreta disposal. Lack of access to these most basic services
necessary to sustain life lies at the root of many of Africa's current
health, environmental, social, economic and political problems."[8] Not
only water, but sanitary arrangements in general are at issue. Puerperal sepsis (childbed fever) is caused by bacteria (beta-hemolytic
streptococci) that are easily killed by penicillin, a drug that costs less
than a penny a dose in Africa. Nevertheless, mortality from puerperal
sepsis in sub-Saharan Africa is as high as it was in the slums of Glas-
gow or the backwoods of New England before Dr. Holmes pinned
down how the malady is spread: skin-to-hand contact—just like Ebola
virus.[9] I have no clear notion, nor had the UN conferees on AIDS, as to
how one gets clean water flowing and hands washed over that great
continent, but until that day arrives, the bill of mortality should
remain unaffected. Those among us who despair of other "lesser
breeds without the law" might do well to recall that clean, publicly
supplied water did not reach New York or London until the 1840s.
Holmes's friend Emerson joined Carlyle in admiring Sir Edwin Chadwick, the dean of English sanitarians "who proposes to provide every

house in London with pure water, sixty gallons to every head, at a penny a week; in the decay and downfall of all religions, Carlyle thinks that the only religious act which a man nowadays can securely perform is to wash himself well."[10] Since our governors are now determined to provide faith-based social services, I'm ready to sign on for those sixty gallons on behalf of godliness. It's little enough in the year of the genome.

October 2, 2001

Postscript: Bellevue in the Terror

TERROR ON THE STREET

IT'S NOW three weeks since the al-Qaida hijackers toppled the Twin Towers and killed over four thousand New Yorkers. Here at Bellevue, largest of our city hospitals and two miles from "Ground Zero," the staff was prepared for an avalanche of casualties. Operating suites were cleared, respiratory teams assembled, ICUs at the ready. In the event, only three hundred patients were seen, fifty admitted, one died. For two days the suites remained empty but the air outside was filled with dust and the smell of something burnt. The major trauma here was psychic: attendings, house staff, students, nurses, aides—everyone— was deeply depressed that there were no survivors, that Bellevue could not do more. The scars remain. They've been building a new, block-long outpatient facility in front of the old hospital and the excavation site is bounded by a hoarding of blue-painted plywood, five feet high. Since September 11 the entire structure has become covered by fliers of the missing: touching wedding pictures, enlarged vacation snaps, passport photos, ID card offprints. It's a reminder of the first few hours after the attack when desperate families scurried from one emergency room to another among downtown hospitals in search of news. The names and faces on the fliers, now protected from the weather by shrouds of plastic, illustrate what former mayor David Dinkins called the "gorgeous mosaic" of our city, as do the grim passersby who stop and stare at this makeshift memorial. At its base, votive candles burn amidst a heap of floral bouquets. More fliers, flags, buntings, fire-engine logos, and police badge decals are plastered over pillars at the hospital entrance. The flowers and candles are refreshed almost daily by friends and family who return there to search for someone with news of the missing.

In fact, many of the missing *are* right here. A three block area around the Medical Examiner's office at our northern boundary is roped off and guarded by state troopers and city cops. A large parking lot has been newly fenced, tented over and is now filled with eighteen huge refrigerated trucks that throb day and night. It's where the dead and/or their parts rest until the MEs have done their job. The trucks are filled and emptied during the day as plastic bags move from ambulance to tented worktables to the MEs receiving bay. The three blocks swarm with troopers, more cops, and platoons of scrub-suited medical examiners and their assistants who work around the clock; outdoor canteens are staffed by the Salvation Army and the Red Cross. State troopers forbid photographs of the fetching and carrying, the grim worktables, the makeshift fences hung with black crepe. Pulled up on the sidewalk are the official vans of police, crime scene investigators, forensic specialists, coroners, and pathologists who have come to Manhattan from all over the country. Next to a mud-covered van of the "Sublette County Wyoming Deputy Coroner" is parked a dusty SUV of a "Crime Scene Investigations Unit" from Suffolk County, New York. A coroner's unit from Minnesota unloads its gear alongside a sheriff's cruiser from Coral Gables.

Only 321 people have been identified so far. It will take that volunteer army months to work out the details of the grisly Who's Who in the trucks and it will take many more months for the DNA results to come out of the sequencing factories of Millennium and Celera. But the scene at Thirtieth Street and the East River looks less like the site of a criminal investigation than the aftermath of a cruel battle, a Matthew Brady photograph of Antietam. It may not be the last battle. As I watch a diverse group of third-year medical students in scrubs and whites threading their way past the police barricade to the ER, it's hard to dismiss this week's warning issued by al-Qaida's chief military commander, Naseer Ahmed Mujahed, "Wherever there are Americans and Jews, they will be targeted."[1] We may see more of what Louis MacNeice called

> The little sardine men crammed in a monster toy
> Who tilt their aggregate beast against our crumbling Troy.[2]

SPORES FROM THE SKY

Many Americans feared that they might become targets of weapons deadlier than hijacked airliners, with airborne anthrax at the top of their worry list: "It would require just a small private plane, not a hijacked commercial jetliner.[3] A helper could casually dump a bag of powdery bacterial spores while in flight, rather than having to over-power a planeload of passengers. And the team could land and be home in time for dinner instead of ending it all in a suicidal inferno."[4] And sure enough, a week after the *Washington Post* worried about a "small private plane," the FBI banned crop-duster aircraft from the vicinity of centers of population. The feds had learned that one of the hijackers in the World Trade Center massacre, Mohamed Atta, had inquired about technical specifications of crop dusters in Florida. In Minnesota, another suspect had crop-dusting manuals on him when he was taken into custody for passport violations.[5]

Inhalational anthrax is serious stuff, indeed.[6] A World Health Orga-nization report estimated that fifty kilograms of aerosolized *Bacillus anthracis* spores, if dispersed by an airplane two kilometers upwind of a population center of five hundred thousand unprotected people in ideal meteorologic conditions, would travel farther than 20 km and kill ninety-five thousand people. But, experts point out that ideal conditions are unlikely ever to apply, citing as evidence careful studies of a single, accidental outbreak of inhalational anthrax loosed from a secret Russian biological warfare facility in 1979.[7] On April 2, 1979, the epidemic swept through Sverdlovsk, killing at least sixty-four people and sickening scores of others. Officials at first blamed the outbreak on tainted black market meat, i.e. *intestinal* anthrax. But soon Soviet defectors leaked news that anthrax spores escaped into the air after an accident on April 2 in Compound 19 of the factory. Western scientists became concerned and in 1992, Boris Yeltsin, new to power, permitted a team led by famed Harvard biochemist Matthew Meselson to look into the facts of the Sverdlovsk epidemic *in situ*. Jeanne Guillemin, who is Meselson's wife and a social anthropologist at Boston College, accompanied the team. Her masterful account of the investigation, "Anthrax, The Investigation of a Deadly Outbreak," has become a manual of inhalational anthrax for those who fear it today.

When the Americans arrived it was still unclear whether the anthrax epidemic was due to intestinal or inhalational anthrax. Guillemin herself went door-to-door to interview survivors, while the rest of the team, using weather reports, travel logs, and factory records found that on April 2, 1979, the wind blew anthrax spores in a single deadly fan originating from the secret plant. The map of where the victims were at the time of the accident, writes Guillemin, gives "a graphic image of a specific event in time, and not just an ordinary event, but of violent, collective death. . . . All the data—from interviews, documents, lists, autopsies, and wind reports—now fit, like pieces of a puzzle. What we know proves a lethal plume of anthrax came from Compound 19."[8] Pertinent to our concerns today is the team's conclusion that the city of Sverdlovsk was lucky. Obedient to prevailing winds, the aerosol of anthrax spores had traveled in an elongated plume that infected human victims southeast of the facility to dissipate over the less populated countryside where it hit only livestock. Had the prevailing winds on April 2 blown north over the densely populated city center, the death toll would have been far greater. Chance ruled the spores.

HOW THE HIJACKER KILLS

Guilleman tells how two courageous Russian pathologists on the scene, Drs. Faina Abramova and Lev Grinberg, made hasty duplicates of their autopsy notes as the KGB confiscated the originals. Their reports were not published in complete detail until this May, when the final paraffin-block tissue samples and autopsy record were reconciled.[9] Death entered by the airways. Lymph nodes in the windpipe and mediastinum were first affected. Lung function was compromised by mediastinal expansion and water in the lungs, followed by blood-borne and backward spread of the anthrax bacilli via lymph vessels to the lungs themselves where they produced pneumonia. The bacteria spread throughout the system where they produced blood vessel inflammation in the brain and gut, with severe bleeding and tissue swelling. The severe terminal hemorrhagic episodes resembled those seen in hemorrhagic fevers like dengue or Ebola. Each—save the last—of these features is consistent with experimental models of anthrax in which the inhaled spores are picked up by mediastinal lymph nodes, followed by bacterial germination, replication, and systemic spread. Edema and cell death are caused by two defined components of the toxin, edema factor and lethal factor.[10]

Anthrax toxin is a deadly mix of three foreign proteins that hijack the machinery of our own white cells to destroy us. The proteins are known as protective antigen (PA), lethal factor (LF), and edema factor (EF). After the bugs release them as single proteins, PA, LF, and EF undergo self-assembly on the surface of the host's own cells. PA first binds to specific receptors and becomes activated by one of our macrophage's abundant surface proteases. The activated form of PA now forms ring-shaped suction cups that bind LF and EF with high affinity. The resulting cell-associated complexes become invaginated, taken up, and trafficked to an acid vacuole, an endosome. Activated PA inserts itself into the endosomal membrane and injects the toxic EF and LF back into the free cell sap.[11] If this pathway sounds familiar, it should. It's the one our own cells use for the routine domestic traffic of the major histocompatibility complex class I molecules, molecules that flag our identity in tissue transplantation. One might say that the anthrax hijackers have slipped through security with fake IDs.[12]

CAPTOPRIL FOR ANTHRAX

Now for news from the lab. My friends who've worked in biodefense tell me that it's hard to cause major mischief in urban populations with simple aerosols that can be diverted by tall buildings. Early detection and public health surveillance are critical: we've learned from lab studies that inhalational anthrax in its early stages can respond well to simple antibiotics. Current recommendations include penicillin G, 4 million units every 4 hours ciprofloxacin, 400 mg every 12 hours; or doxycycline, 100 mg every 12 hours.[13] But, thanks to modern molecular biology there's other help on the way. Edema factor has been identified as an adenylate cyclase, an enzyme that disarms our white cells from within by raising their cellular cyclic AMP. It's been known for decades that methlyxanthines such as caffeine, theophylline, or theobromine enhance this dampening effect of cyclic AMP.[14] Conclusion: no coffee, tea, or Coca-Cola when anthrax is in the air.

There's more: it turns out that lethal factor—for reasons of evolution that are murky at best—looks for all the world like a curious enzyme involved in the generation of leukotrienes, second cousins to inflammation-producing prostaglandins, some of which also raise cyclic AMP. The enzyme is called LTA4 hydrolase, a two-headed protein, one end of which is a zinc-containing metallopeptidase. Lethal factor, which counts among its intracellular targets a number of signaling kinases, resembles LTA4 hydrolase in its susceptibility to a number of well-known, well-studied metallopeptidase inhibitors, such as the widely used antihypertensive agent captopril![15] Since captopril protects animal and human cells against anthrax toxin, we have on hand today a promising defense against lethal factor and its fellow hijackers. Since a number of captopril-like drugs have already been developed, we can soon look forward to a tighter homeland defense.

Anthrax has been around as scourge of man and beast since prehistory. But it wasn't until 1875 that Robert Koch discovered B. anthracis and began the bacteriological revolution.[16] Since then, thanks to rational—and international—science we've learned where the bacterium lives, where it hides, how it kills, and how to eradicate it. That's a pretty good model for coping with the ancient scourge of zealotry that is lapping at our shores. Auden counted the cost of such a battle the last time democracy was in play:

> However we decide to act
> Decision must accept the fact
> That the machine has now destroyed
> The local customs we enjoyed.[17]

[Author's note: This entry was written two days before the first death from inhalation anthrax in the United States. The World Trade Center death toll as of January 2002 is approximately three thousand.]

August 6, 2001

Ever since Galileo:

Cloning Loses in the House

EMBRYO FEARS

THIS WEEK the U.S. House of Representatives voted 265 to 162 to ban human cloning not only for reproductive purposes, but also for the treatment of disease and for research.[1] After a tangled debate that dealt with difficult issues of ethics, science, and religion in soundbites appropriate to the evening news, the House rejected a compromise bill that would have permitted therapeutic cloning to proceed. "I think the House spoke very, very loudly today that this is morally and ethically inappropriate," said Representative Dave Weldon, (R) Florida, the bill's chief sponsor. "It clearly sends a message that there is a place we don't want to go, and that is the manufacture of scientific embryos for research."[2] Opponents of cloning—and of stem cell research in general—were overjoyed. Douglas Johnson, speaking for the National Right-to-Life Committee crowed that "The House has acted to block the creation of embryo farms." A majority of the House had been swayed by arguments that confused the clonal expansion of nuclear transplants in a dish with the implantation of a fertilized ovum into a uterus. Therapeutic cloning [see September 5, 2000], the option turned down by the House, involves placing the nucleus of a somatic cell into an anucleate human egg (oocyte), activating the now diploid cell by physico-chemical means rather than sperm (parthenogenesis), and differentiating it into a functioning clone of cells in vitro. Peter Deutsch, (D) Florida, told the house that a therapeutic clone of a nuclear transplant "is not an embryo. It is not creating life by any definition of creating life." Mr. Weldon replied, "That's like saying Dolly is not alive." Weldon is wrong. Dolly may have begun as a nuclear transplant in a dish, but like all of us, she came alive in a womb.

Never mind those "embryo farms." English, French, and American dictionaries agree on the definition of an "embryo," it requires a womb.[3] Lawmakers and reporters alike are therefore wrong to worry that "scientists would create embryos that could be used to treat disease."[4] The diploid blobs of cells used in therapeutic cloning are like the embryoid bodies produced in one version of stem cell therapy.[5] Neither embryoid bodies nor nuclear transplants are embryos, as Rep. Weldon might find in an excellent report on stem cell science prepared for his own head of Health, Education, and Welfare (HEW) on July 18, 2001.[6] The National Institutes of Health (NIH) document also makes the point we're already doing therapeutic stem cell cloning. Every bone marrow transplant contains diploid stromal stem cells that will expand clonally in the recipient.

IMMACULATE CONCEPTIONS

Many of the legislators in the House were no doubt swayed by opinions spelled out in a very different kind of monograph. Issued last summer by Vatican scholars, it reasserted that life begins when sperm meets egg, preferably in vivo.[7] The document, which in high scholastic dudgeon condemned cloning and/or stem cell research on doctrinal grounds, was cited by the pope when President Bush met him on July 23.[8] The pontiff specifically declared that the creation of human embryos for research purposes was an "evil akin to euthanasia and infanticide." His supporters in the United States went further, warning Congress against creating "a ghoulish industry."[9]

News of the congressional ban was received with disappointment here at the Marine Biological Laboratory at Woods Hole, where reproductive biology has flourished for over a century. In 1900 Jacques Loeb was accused of transgressing the natural order when he produced viable sea urchin larvae by means of parthenogenesis.[10] Religious opinion was divided at the time; some argued that parthenogenesis proved the possibility of immaculate conception, others were appalled that "It would soon be possible to raise domestic animals and children born without help of a male."[11] Debates followed over the morality of fatherless babies, scholastic notions of "natural law" and "arguments from design" were used against Loeb, the champion of reductionist science. Replying to one such critic, the Jesuit Father E. Wassman, Loeb wrote "Such scholastic discussions are very serviceable for the defense of a dogma or an opinion. . . . But one cannot overlook the fact that the steady progress of science dates from the day when Galileo freed science from the yoke of this sterile scholastic method" (read: authority).[12] To calm the waters, Loeb's colleague, F. R. Lillie, assured a local newspaper in 1900 that the creation of vertebrates by parthenogenesis was a very long way off, indeed. Why, it would be like finding the North Pole, he said.[13] Admiral Peary found the North Pole nine years later.

PINCUS REDUX

By 1935, Gregory Goodwin Pincus (1903–1967), best known as "The Father of the Pill," had worked out techniques for fertilizing mammalian eggs in vitro. An untenured assistant professor at Harvard, he also reported that he had activated rabbit eggs with sperm in vitro, reimplanted them in female rabbits, and obtained viable fetuses.[14] In the summer of 1936 at Woods Hole, he told future Nobel laureate Herman Muller that he had activated "other" mammalian eggs parthenogenetically.[15] Of course, the problem with parthenogenesis is that the activated, reimplanted eggs become haploid fetuses (twenty-three instead of forty-six chromosomes in humans). Pincus knew that haploid sea urchins don't develop much past the larval stage, and that haploid mammalian embryos have an even harder time of it in utero. Nevertheless, "Presumably normal mammalian embryos might develop if a diploid nucleus could be induced to form."[16] That didn't happen until Dolly, and her nucleus came from a diploid cell. Pincus knew in 1935 what Congressman Weldon doesn't know in 2001, that you need a womb to make an embryo. Pincus understood that "the differentiating embryo is dependent upon a uterine environment the optimum development of which involves a fairly definite time schedule."[17] He again raised the banner of reductionist science: "Careful investigation of the ovum itself and its homeostatic environment is made possible by the various explantation and transplantation techniques."[18]

"Explantation" and "transplantation," "parthenogenesis" and "fatherless embryos"—charged terms that make headlines today made even more sensational headlines in the 1930s. The *New York Times* worried that Pincus's work would lead to embryo farms and human cloning à la Aldous Huxley's *Brave New World*.[19] Darker themes were sounded that smacked of Father Coughlin and anti-Semitic nativism. *Collier's* magazine published an attack on Pincus that featured an unflattering photo of this "Jersey native" as a cigarette-dangling mad scientist, holding a "fatherless" rabbit under his arm. The writer J. D. Ratliff traced Pincus's "goofy" experiments on parthenogenesis and in vitro fertilization to the work of "Jacques Loeb, the hugely famous Portugese Jew." As in Loeb's day, this fan of natural law worried that with parthenogenesis "Man's value would shrink, the mythical land of the Amazons would then come to life. A world where woman would be self-sufficient; man's value precisely zero."[20] Pincus was denied tenure at Harvard.

We've now had a replay of some of those darker notes in the Congress of the United States, but whatever happens next, I'm glad they haven't turned the clock back on in vitro fertilization, nor on Pincus and his pill. Indeed, the lawmakers are unlikely to stop scientists—in God's country or elsewhere—from working on "the various explantation and transplantation techniques" required to treat the sick with healthy clones. Not surprisingly, workers in Japan have already figured out an explantation technique for producing "parthenogenones," i.e., activated, haploid, human oocytes. Pincus redux in Tokushima.[21]

THE CHECK IS ON THE MALE

Sheldon Segal, of New York's Population Council and the Marine Biological Laboratory, Woods Hole, Massachusetts, suggested to me that one way to bypass doctrinal objections to working with discarded, sperm-activated embryos is to study blastocysts derived from "parthenogenones." The procedure works in goats.[22] But, is haploid life possible? Diploidy (as in nuclear transplants) has always been considered a superior state because if one copy of a gene gets mutated, a second copy of a gene can compensate. That dogma fell last month when Dutch scientists described the first metazoan species that reproduces by parthenogenesis—*Brevipalpus phoenicis*, the "false spider mite" found on a Brazilian coffee plantation, has dietary habits that resemble those of John Wayne in a Western saloon: the mite eats, shoots, and leaves.[23] Unlike barflies, however, the mite colonies are composed entirely of haploid females with one rather than two (if diploid) chromosomes. Even more surprising, this "reproductive anomaly" is due to a novel intracellular bacterium that causes feminization of genetic males. As proof, when the female mites were treated with antibiotics, close to half their offspring became male. The bacteria produced a substance that was able to check male differentiation and to insure that the females produce infected female progeny by parthenogenesis.

Diploidy does not rule the roost everywhere in nature. Haplodiploidy (haploid males, diploid females) has turned up at least seventeen times in evolution. Since the mid-70s we've known that this odd chromosomal sorting is mediated by a rickettsial symbiont, Wolbachia.[24] Wolbachia may be the most common infectious bacterium on earth. No vertebrates are known to carry the microbe, but infection is common in many invertebrates: fruit flies, spiders, shrimp, lobsters, and round worms. Rickettsia-like, maternally inherited bacteria have been shown to be involved in a variety of alterations of invertebrate sexuality, such as female-biased sex ratios, parthenogenesis, and sterility of crosses either between infected males and uninfected females or between infected individuals (cytoplasmic incompatibility). This jamming of the host's reproductive machinery insures the passing down of Wolbachia to future generations because the bacteria are passed on vertically only from mothers to daughters. Males are a dead end for intracellular bacteria: the infected females win because infected male sperm are killed by a toxin made by the microbe.[25]

POTOMAC RIVER BLINDNESS

Research on parthenogenesis, haploidy, gender determination—all the science that fundamentalists believe to violate natural law—has had another surprising fallout.[26] When Wolbachia infect filarial worms, such as those that cause elephantiasis or river blindness, they reach a symbiotic state, in which the filaria cannot live in their human host unless the worm is in turn infected by Wolbachia. Taking advantage of that knowledge, German scientists have treated patients suffering from river blindness (onchocerciasis) with the antibiotic doxycycline, and lo and behold, the patients improved as their parasites were cured of *their* parasites.[27] Moreover, since tissue injury in river blindness is caused by neutrophils attracted by the worms, it was good news that "worm nodules from untreated onchocerciasis patients displayed a strong neutrophil infiltrate adjacent to the live adult worms. In contrast, in patients treated with doxycycline . . . neutrophil accumulation around live adult filariae was drastically reduced."[28] Good news for the hundreds of thousands suffering from filarial diseases the world over.

Loeb had it right, the steady progress of science dates from the day when Galileo freed it from authority. When lawmakers ban research in any area of science, they can have no inkling of how many roads they've closed, of how many of their fellow creatures their irrational fears will have condemned to blindness. With therapeutic cloning, we're not out to create a "Brave New World," but a healthy one.

July 9, 2001

Nobody Loses Sleep over
Sleeping Sickness

THIS WEEK the gap widened between the medical needs of developed and undeveloped countries. In Louisville, a battery-operated, artificial heart (cost approximately seventy-five thousand dollars) was successfully implanted in the chest of a fifty-year-old diabetic man fated to die of congestive heart failure.[1] In Washington, sixty-year-old Vice-President Dick Cheney had a battery-operated pacemaker/defibrillator (approximately thirty thousand dollars) implanted in his chest to protect him from a fatal run of ventricular tachycardia.[2] A wag suggested that this put George W. Bush one irregular heartbeat away from the presidency. At Woods Hole, Massachusetts, molecular parasitologist Keith Gull of Manchester announced that DNA microarray technology (approximately twenty thousand dollars) had helped him figure out how *Trypanosoma brucei* cause sleeping sickness. In Freetown, Sierra Leone, "The Worst Place on Earth," where thousands of children have been impressed into militias and mutilated by machetes in the course of a nine-year-long civil war, the World Health Organization and Médecins Sans Frontières mounted a campaign against sleeping sickness.[3,4] They were aided by eflornithine, a cosmetic drug that removes unwanted facial hair in Western women.[5] It was the end of a long campaign by MSF to bring back the drug that was in danger of disappearing because its manufacturer found it unprofitable. Despite five hundred thousand victims and 55 million at risk, MSF complained that, unlike AIDS, "Nobody loses sleep over sleeping sickness."[6]

TSETSE AND TRYPANOSOMES

At Woods Hole, Keith Gull reminded a packed audience at the Marine Biological Laboratory of the rich history of sleeping sickness. Microbe hunters had known since the 1870s that trypanosomes caused diseases of animals (e.g., nagana in cattle), but the parasites were not considered a danger to humans. At the turn of the century, however, Sir David Bruce (1855–1931), who also discovered the agent of brucellosis, worked out the life cycle of the parasite that bears his name, found that it was transmitted by the tsetse fly, identified its animal reservoir in African wild game and established that it was the cause of African sleeping sickness.[7] Another pioneer, Sir Aldo Castellani (1877–1971), a British-Italian bacteriologist, who also discovered the spirochete of yaws, pinned the matter down by finding the trypanosome in the spinal fluid of a patient with the disease. Yet the Nobel Prize for trypanosome research in 1907 went to Charles Louis Alphonse Laveran (1845–1922) of Paris, who also discovered the various blood forms of the malaria parasite. His Nobel citation honored him for linking the geography of insect vectors to animal and human disease:

> All these infections are caused by corkscrew-shaped micro-parasites, called trypanosomes, and are transmitted to animals by various types of biting flies. However important these diseases may be to Man from the point of view of commerce and nutrition, yet, among all the trypanosomiases, the endemic disease generally known as "sleeping-sickness" takes precedence from the medical point of view.[8]

In Africa, the geographic distribution of flies, trypanosomes, and the animal reservoir has been unchanged for a century. A map Laveran published of tsetse fly–borne diseases in 1912 looks for all the word like the map of tsetse fly distribution and sleeping sickness in July of 2001. Outbreaks of sleeping sickness come and go in epidemic waves as the host-parasite balance shifts. Right now the hosts are losing and sleeping sickness is resurgent. In some villages in Angola, in the Democratic Republic of Congo (DRC), and in the southern Sudan, it is the biggest cause of death, ahead of AIDS.[9]

We now recognize three subspecies of *Trypanosoma brucei*: (a) *T.b. brucei* infects animals but not humans, because our cells produce a unique "lytic factor" for this subspecies, (b) *T.b. gambiense* causes chronic, or West African, sleeping sickness, (c) *T.b. rhodesiense* produces the acute or East African form of the illness, a subspecies that was not identified until 1910.[10,11] Unfortunately, only *T.b. gambiense*—the kind found in the DRC and in Sierra Leone—is sensitive to eflornithine, which has been hailed by the WHO as "the resurrection drug."[12]

SUNDAY IN THE PARK WITH GENES

Keith Gull explained that vaccines are unlikely to work against *T.brucei* because the parasite dodges the host's immune system by "antigenic variation" of its variant-specific surface glycoprotein (VSG) coat. Its surface is packed with 10^7 copies of these deceptive glycosylphosphatidylinositol (lipid)-anchored VSGs. Epidemics result when the bulk of the parasite population has adopted a new surface coat that the host population hasn't seen before.[13] Instead of vaccines, Gull argued, we need drugs directed against molecular targets that the parasite does not share with the host. Happily, some of these have already been identified and others are popping up on DNA microchips.[14] *T.brucei* doesn't express its genes quite the way we do. The genes encoding the VSGs of infective *T.brucei* are located at telomeres and their expression is characterized by what George Cross has called "the heresy of RNA editing," including the novel trick of double-stranded RNA-induced RNA degradation or interference.[15,16] Since trypanosome genes contain no introns they use RNA trans splicing which adds an identical leader sequence to all trypanosome mRNAs. Those are the mRNAs picked up on the little shiny dots of DNA microarrays, the patterns of which change as the surface coats change. Pictures of these thousand points of color reminded Gull of Georges Seurat's pointilliste masterpiece *A Sunday at la Grande Jatte.*

COSMETIC PHILANTHROPY

But the treatment of sleeping sickness is no picnic in the park. The drugs now in use, eflornithine excepted, are almost as noxious as the disease itself. Melarsoprol, discovered in 1949, is the last arsenic derivative still used in trypanosomiasis, and is the only one that can cross the blood-brain barrier. Its side effects are awful. "In the field we are forced to use melarsoprol and when we put it into people's veins, they often scream," said Daniel Berman of MSF.[17] It kills 3 to 10 percent of patients, and others suffer brain damage. Far less toxic is eflornithine, which was discovered in 1980 by Dr. Cyrus Bacchi of Pace University, as part of a study of polyamine metabolism in trypanosomes.[18] Eflornithine is a site-specific inhibitor of ornithine decarboxylase, a rate-limiting step in trypanosomal cell division. Eflornithine has been repeatedly found effective against *T.b. gambiense*, but was "unaffordable" by authorities in the DRC, Angola, or Sierra Leone, the treasuries of which were depleted by the cost of fratricide.[19] In consequence, manufacture of bulk material ceased in 1995. After agitation by MSF, Hoechst Marion Roussel (now Aventis) granted WHO the production rights for eflornithine in December 1999, hoping to interest a third-party manufacturer.

Matters changed when it was found that ornithine decarboxylase was also the rate limiting enzyme in the development of hair follicles and that topical eflornithine worked to remove pesky chin whiskers from women's faces.[20] Since cosmetics are very big business, indeed, it became worthwhile to manufacture the drug again. Made in bulk, unit costs of the material drop sharply.[21] In March of this year, Aventis together with Bristol-Myers Squibb and Gillette, who had gotten the eflornithine cream Vaniqa approved by the FDA, agreed to donate bulk material for sixty thousand doses of eflornithine to MSF by June 2001. This supply will last for about a year.[22]

THE SHIELD IS NOBLER THAN THE SPEAR

Alas, even with eflornithine available, even with WHO and Médecins Sans Frontières manning the battle stations, I'm afraid that the 55 million at risk for sleeping sickness are likely to remain at risk. Part of the problem is the general lack of sanitation, part the absence of fly control. But the greatest part is the terrible disruption of civil and tribal war. James Traub reports in the *New York Review of Books* that

> Now Sierra Leone is known as the place where its rebels chop off peoples' hands and feet, rape little girls and old women, press-gang children into combat and use civilians as human shields.... On Pademba Road, a main thoroughfare where part of the UN headquarters is located, you have to hop deftly over gaps in the paving stones of the sidewalk in order not to fall into the sewer.[23]

Meanwhile, medical resources have been diverted to cleaning up the human consequences of terror. In Sierra Leone, as in the DRC, Sudan, Rwanda, Uganda, and Angola, people are losing sleep over single- and double-arm mutilation.[24] In June, some 136 children on whose faces and bodies rebels carved their groups' acronyms during Sierra Leone's civil war were to have them removed by two visiting plastic surgeons from the International Medical Corps.[25] Sleeping sickness is resurgent, and not only because of tribal warfare and the diversion of medical resources. The sewers leak, flies breed in the water, the sick go begging in the gutter; average life expectancy in Sierra Leone is forty-two years. More Africans die of febrile, diarrheal diseases than of either AIDS or sleeping sickness and those fevers and dysenteries come from contaminated drinking water.[26] Dr. Oliver Wendell Holmes wasn't describing Sierra Leone, but the United States a century and a half ago when he wrote:

> Look at the annual reports of the deaths in any of our great cities, and see how their regularity approaches the uniformity of the tides, and their variations keep pace with those of the seasons. The inundations of the Nile are not more certainly to be predicted than the vast wave of infantile disease which flows in upon all our great cities with the growing heats of July—than the fevers and dysenteries which visit our rural districts in the months of the falling leaf.[27]

We've learned that the sanitary revolution was the sine qua non of medical progress in the West and I daresay that Africa might be better served by gifts from Hygea than from Aventis. We owe to the sanitarians of the nineteenth century—among them Holmes and Howe in America, Chadwick and Nightingale in England, Proust (*père*) and Zola (*père*) in France—the increase in life expectancy in the West from circa forty years in 1840 to sixty by 1940. Holmes's anthem bears repeating:

> To guard is better than to heal
> The shield is nobler than the spear[28]

Without the shield of sanitation, we couldn't have forged all those spears of medicine in 2001: the defibrillators, the artificial heart pumping in Louisville, yes, even the depilatories. Clean the water, kill the flies, and all the rest will follow. The sleepers will wake.

June 12, 2001
Arsenic Redux

ARSENIC is in the news: it may or may not be contaminating our drinking water, it may or may not have murdered Napoleon, but it has been approved by the FDA for treating promyelocytic leukemia where it may or may not team up with wonder drug Gleevec to become the oldest new cancer therapy on record.

THE ARSENIC PRESIDENT

This week, when President Bush travels to Europe, he will meet folks who know him as "the president who put arsenic back in the drinking water."[1] Arsenic may be the least of his problems; his own officials confess that "the common European conception is of a shallow, arrogant, gun-toting, abortion-hating, Christian fundamentalist Texan buffoon.[2] On this side of the Atlantic, calmer views prevail, but to American environmentalists President Bush is still "the Arsenic President."[3] What the *Wall Street Journal* called "The Arsenic Ruckus" became an issue in March, when Christie Whitman, new head of the Environmental Protection Agency, delayed a last-minute rule issued by the Clinton White House to reduce arsenic in drinking water from fifty parts per billion to ten parts per billion (= 50 to 10 μg/L).[4] Whitman said that a further review was needed and called for a new study to focus on health concerns of levels between three and twenty parts per billion. She called on the review to be completed by February 22, 2002, but hoped the process could be finished by the end of 2001. The Natural Resources Defense Council (NRDC) countered with a lawsuit, calling suspension of the 10 μg/L limit "scientifically unwarranted and illegal." The council pointed out that 10 μg/L is the standard set by the World Health Organization and the European Union. "Unfortunately, President Bush apparently won't listen to reason, scientific evidence, or the will of the American public, so NRDC is left no choice but to sue," Tad Olson, a spokesman for the NRDC, said. Olson believes that the EPA should have lowered the standard for arsenic in drinking water to three parts per billion. He explained that the Clinton EPA had set a 5 μg/L standard last June, but increased it to 10 μg/L in response to industry pressure.[5] Whitman turned the problem back to the Commission on Life Sciences of the National Academy of Sciences, a group that had issued a definitive, if cautious, report entitled "Arsenic in Drinking Water."[6]

In its 1999 report, the commission presented clear evidence that arsenic is a human carcinogen and noted that since the 1940s, the conventional maximum contaminant level (MCL) for arsenic in drinking water had been 50 μg/L (50 micrograms per liter). But a thorough review of all the data persuaded the commission that an "MCL for arsenic in drinking water of 50 μg/L does not achieve EPA's goal for public health protection and, therefore, requires downward revision as promptly as possible." Absent exact knowledge, no lower limit was set. Whitman delayed implementation of this report, claimed that further review was needed, and called on the same commission—plus or minus a few members—to "focus on health concerns of levels between three and twenty parts per billion." This week, the *Wall Street Journal* and the Bush regime found support on National Public Radio. NPR's Larry Abramson interviewed several administrators of local water districts. One of these chaps, intent on balancing the county budget, was convinced that the cost of decreasing arsenic content probably out-weighs its benefits. How much is a life worth? No answer from the water works.[7]

Tri- or pentavalent Arsenic (As) is a rather unique carcinogen; it cannot produce internal cancer in laboratory animals. However, at doses between 10 and 100 μg/L it produces carcinogen-like effects in culture and mimics the progression of skin cancer from keratoses to atypical keratoses to squamous cell carcinomas as in human arsenic poisoning.[8] Humans exposed to levels of about 250 μg/L will clearly develop kidney, liver, and other internal tumors in addition to the characteristic skin lesions. This toxic level was first established in a small cohort of British beer drinkers over a century ago and has been disputed since.[9] Data at lower levels are inconclusive and drawn from equally distinct populations in Taiwan, Bangladesh, and Chile. But if 250 μg/L is bad news, indeed, I'd go along with the Natural Resources Defense Council and vote for an MCL of 2.5 μg/L. Tops. That's cut-ting our risk by two log orders. But I'm afraid that the level of arsenic we'll get in our drinking water will reflect a preference for the pocket-book over public health. Stay tuned for Whitman's call on arsenic after you've gotten your tax refund.

THE ARSENIC EMPEROR

The French had arsenic on their minds in another connection. Medico-legal experts met at Strasbourg last week to decide whether the fifty-one-year-old Napoleon died of arsenic poisoning on May 5, 1821, on the South Atlantic island of St. Helena. Accused was Count Charles de Montholon, a companion in exile of the emperor, who was jealous of Napoleon for his passionate affair with the count's wife, Albine. Mon-tholon is said to have been abetted, and well rewarded for his efforts, by the emperor's British captors and the restored Bourbon monarchy. Both parties were anxious to avoid a replay of Napoleon's hundred-day return from his earlier exile at Elba.[10] The story has been around for over forty years, first proposed by a Swedish dentist, Sten Forshufvud, and later elaborated by Ben Weider, a fitness equipment magnate and longtime student of Napoleon who runs the Montreal-based International Napoleonic Society.[11,12] The clinical record—and an autopsy report without microscopic data—would fit either the standard diagnosis of gastric cancer with subacute perforation or the Forshufvud/ Weider notion of "chronic arsenic poisoning with a terminal overdose of calomel in barley water (sirop d'orgeat.)." The kicker in this story is the arsenic in the emperor's hair. Five samples of Napoleon's hair have been analyzed by the best of modern techniques: by spectroscopy at the l'Ecole Polytechnique de Lausanne; by X-ray diffraction at the synchrotron in Grenoble; by the nuclear labs at Harwell in England; and by the toxicology division of the FBI.[13] Weider's relentless pursuit of the theory has led to almost yearly conferences and to a flurry of news stories each time the hairs are jumped through another analytic hoop. At Strasbourg, the experts agreed that the high concentrations of arsenic—between seven and thirty-eight nanograms/mg hair (<1 ng/mg normal)—were compatible with chronic arsenic poisoning. At a similar pow-wow held in the French senate last year, Weider announced: "Both the FBI and Scotland Yard, confronted with the results of these tests, have said that if they came across similar results in the case of a recent victim, they would have no hesitation at all in opening a murder inquiry."[14]

Other experts demurred, arguing that one cannot decide whether the arsenic came from ingestion or external sources. Arsenic might have accumulated thanks to the emperor's love of shellfish or proprietary "tonics"; another theory has it that arsenic dropped into his aperitifs from peeling wallpaper that contained arsenic-based dyes.[15] In fact, since arsenic is a traditional fixative in embalming and taxidermy, it might have been absorbed from fluids in which the hairs were originally preserved.[16] "It would be stupid of me to say poisoning was impossible," said journalist Jean-Paul Kauffman. "But I think Napoleon was poisoned by the ghosts of his past glory rather than by [chemicals]. And his legend needs enigmas, not certainties."[17] But I've got a hunch Weider is right. The finger points to perfidious Albion, whether or not a jealous count spiked the wine. Napoleon's anguished testament of April 19, 1821, tells the tale: "et moi, mourant sur cet affreu rocher, prive des mien et manuant de tout, je legue l'opprobre et l'horreur de ma mort a la famille régnate d'Angleterre" (and me, dying on this frightful rock, given nothing and lacking everything, I leave the shame and horror of my death at the foot of the royal family of England).[18] Heartbreaking, unless you remember those who died on behalf of those "past glories."

THE ARSENIC CURE

Arsenic is not only among the oldest of poisons, it's also among the oldest of cancer remedies.[19] In the eighteenth century, Thomas Fowler developed a solution of arsenic trioxide in potassium bicarbonate (1 percent w/v) which became popular as an oral remedy for dozens of diseases ranging from anemia to zoster. Eventually, arsenic given by mouth became the keystone of antileukemic therapy until it was supplanted by radiation in the early twentieth century. It's back; this time by vein.

Arsenic trioxide was approved by the FDA last November as a treatment for acute promyelocytic leukemia—a rare disease affecting less than two thousand per year.[20] Formulated as Trisenox by Cell Therapeutics of Seattle, arsenic now looks promising in other cancers, including multiple myeloma.[21] This week news came that arsenic acts in concert with Gleevec, the "magic bullet" against chronic myelogenous leukemia.[22] Arsenic trioxide acts on the same faulty circuit in carcinogenesis as Gleevec, albeit at another switch. A group from Tampa showed that arsenic trioxide lowers the levels of BCR-ABL (the Philadelphia chromosome enzyme) in leukemic cells and when it is combined with Gleevec, which directly inhibits that enzyme, it "can potently induce apoptosis and differentiation of BCR-ABL–positive human leukemic cells."[23] Good news for those with chronic myelogenous leukemia, again. Moreover, a group from Dijon found that arsenic works via mitochondrial membranes in many sorts of leukemic cells to bypass their resistance to classical anticancer drugs. Good news for cancer patients with drug-resistant disease.[24]

Arsenic is back thanks to the persistence of Western-oriented Chinese doctors who since the early 1970s have pursued the mode of action of a traditional Chinese medicine, Ai-Lin I, in skin cancer and leukemia.[25] By trial and error, doctors at Harbin Medical University identified arsenic stone powder as the effective component in the remedy. Since oral administration of As_2O_3 was associated with severe side effects to liver and stomach, the drug was further purified for intravenous use. It worked.[26] Dr. Raymond Warrell brought the Chinese formula to Memorial Sloan-Kettering, French and American studies were launched, and the first results from studies in the West were published in the November 5, 1998, issue of the *New England Journal of Medicine*. The title tells it all: "Complete remission after treatment of acute promyelocytic leukemia with arsenic trioxide."[27]

The drug was approved less than two years later and trials in a variety of cancers, with and without conventional chemotherapy, are under way. News that oncologists are about to use arsenic as combined therapy with the newest of anticancer drugs would have pleased Paul Ehrlich who had such high hopes for arsenic-based chemotherapy. He predicted in 1913 that "A further advantage of combined therapy is that when two different medicaments are used resistance of the parasite to a drug, e.g arsenic, is far less likely to arise. . . . Combination therapy is best carried out with therapeutic agents of which each attacks a different chemoreceptor, in accordance with the military maxim: march in detachments, fight as a unit."[28] The maxim could have been Napoleon's.

May 16, 2001

Dr. Baltimore's Magic Bullet

ON MAY 10 the secretary of health and human services, Tommy Thompson, announced that an effective oral treatment for chronic myeloid leukemia (CML) had been approved by the FDA. The drug, Novartis's Gleevec (imatinib mesylate), had breezed through in less than three months; a very fast track, indeed. In the awkward prose of the Bush administration, Thompson declared, "Today I have the privilege of announcing one of those medical breakthroughs that is outstanding." The acting head of the FDA, Commissioner Bernard Schwetz, was less stunned: "Although the long-term benefits of the drug are not yet known, early studies have shown that Gleevec will offer a significant improvement for many patients."[1] The announcements in Washington fell short of spelling out that Gleevec is as close to a magic bullet for cancer as any we've got today. By shorting the molecular circuit gone awry in leukemic, but not normal, cells, the drug can induce remissions in over 90 percent of CML. It can even stop the fatal "blast crises" in their tracks. "It's as if you'd cut the noose around the neck of someone on the gallows," explained a Harvard oncologist.[2] The story of how the drug was developed is a tale of two continents, four decades, dozens of labs, and hundreds of patients. At its center is the best scientist of our day, David Baltimore.

BACK FROM THE GALLOWS

Approval of the drug came a month after full publication of pivotal clinical studies in CML led by Brian J. Druker of Oregon Health Sciences University.[3] While Druker's successful trials with Gleevec were based on the work of academic geneticists and molecular biologists, it came to fruition because of chemists, pharmacologists, and the corporate persistence of the Swiss drug giant, Novartis (a.k.a. Ciba-Geigy/Sandoz). Since CML is a relatively rare malignancy that in this country leads to fewer than five thousand deaths a year, the company's commitment of $800 million to develop Gleevec was one large throw of the dice. It was a gamble based on Paul Ehrlich's charge to the microbe hunters of 1910: "to find chemical substances which have special affinities for pathogenic organisms, which would go, as 'magic bullets' straight to the organisms at which they were aimed."[4] Edward G. Robinson, who played the title role in the 1940 film *Dr. Ehrlich's Magic Bullet* (and who died of cancer), would have bet on the outcome.

Druker's data showed that Gleevec had restored normal blood counts in fifty-three of fifty-four patients with chronic myeloid leukemia in whom standard therapy with interferon alpha had failed.[5] Of the fifty-four patients, twenty-nine showed good cytogenetic responses (substantial loss of the hallmark Philadelphia chromosome) while seven had complete cytogenetic remissions. Druker told reporters that fifty-one of the fifty-three responders were still doing well after about a year on the medicine and that they were expected to remain on the drug indefinitely. Richard S. Kaplan of the National Cancer Institute enthused over the first responses: "A 98 percent response rate to a cancer drug is virtually unprecedented, I don't know that in medicine we're going to very often, if ever, do better than this with a single agent."[6] Trials of the drug in the more deadly blast crises of chronic myeloid leukemia and acute lymphoblastic leukemia with the

Philadelphia chromosome were also convincing. With conventional therapy, the five-year survival rate of those in crisis is only 6 percent.[7] With Gleevec, twenty-one of thirty-eight patients (55 percent) with a myeloid blast crisis responded; indeed four experienced a complete hematologic response. The noose had been cut from their neck, so to speak. Moreover, fourteen of twenty patients with lymphoid blast crisis (70 percent) also had a response. "This is a new concept in cancer drug design," said David A. Scheinberg, chief of the leukemia service at Memorial Sloan-Kettering Cancer Center.[8] Well, not really. Big Pharma from Abbot to Zeneca, biotech firms from Ariad to Vertex are targeting oncogene-directed kinases, and clinical trials by the dozen are in the clinic. I'm confident that the Gleevec story is only the first of its genre, that the Novartis gamble will yield as neat a payoff as Dr. Ehrlich's magic bullet.

THE PHILADELPHIA CHROMOSOME

It all began with Peter Nowell, a cellular pathologist at the University of Pennsylvania who found an abnormal, "minute" chromosome in CML, a disease often attributed to ionizing radiation of any source.[9] Further cytogenetic studies by Janet Rowley (University of Chicago) and others established that the abnormality is present in virtually all cases of CML and is also found in 20 percent of acute lymphoblastic leukemias.[10] It was called the Philadelphia chromosome in honor of Nowell's city. CML is the direct result of cellular miscegenation in which pieces of two chromosomes break off and switch places. In the process, chromosome 22 becomes shorter than normal and chromosome 9 acquires an extra-long end. The truncated chunk of chromosome 22 contains the rogue cancer gene BCR-ABL. The critical observations were made by David Baltimore and his student Oliver Witte, who discovered that the enzyme coded by oncogene BCR-ABL is an aberrant tyrosine kinase, which—like other oncogenes—is a molecular switch irreversibly stuck in the "on" position and therefore unable to control the thermostat for white cell production. By means of gene transfer experiments in vivo and in vitro, they showed that this product of the Philadelphia chromosome is both necessary and sufficient for producing leukemia.[11] Meanwhile, Dr. Alex Matter, head of oncology research at Novartis, had been pursuing the anticancer properties of site-specific inhibitors of BCR-ABL and its sister enzymes in the kinase family.[12] Matter's search for specific inhibitors of the Philadelphia chromosome kinase reached its clinical conclusion in a collaboration with Druker and with Nicholas B. Lydon of Amgen— who had been a collaborator of Druker's since the Oregonian's days at Boston's Dana-Farber.

DRAMATICALLY IMPROVING THE HUMAN CONDITION

Days before the drug was approved by the FDA, five scientists who had made Gleevec possible received what should be the first of many glittering prizes to come. In Boston, Drs. Baltimore, Witte, Druker, Matter, and Lydon received the prestigious 13th annual Warren Alpert Foundation Scientific Prize, awarded by the Harvard Medical School and the Massachusetts Institute of Technology for research that "dramatically improves the human condition."[13] It was the second post-Nobel Prize that Baltimore had earned within the year; last spring he was awarded the National Medal of Science by President Clinton, chiefly for his Nobel-winning work on retroviruses (he has held the post of chairman of the AIDS Vaccine Research Committee of the National Institutes of Health since 1996). This month, a new biography of Baltimore has appeared which documents his extraordinary creativity and the esteem in which he is held by his peers.[14] Baltimore shared the Nobel Prize in 1975 with the late Howard Temin for discovering the enzyme that lets RNA make DNA. This enzyme, reverse transcriptase, became the prime target of anti-HIV drugs, and AZT turned out to be the salvarsan of the class. Baltimore went on to become the founding director of MIT's Whitehead Institute, then president of Rockefeller University and he is now the president of Cal Tech. In the midst of this glittering administrative record, he managed to stay ahead of the curve of modern biology, justifying Crotty's claim that "It is not an exaggeration to say that one could write a pretty decent history of the last 25 years in biology by reviewing Dr. Baltimore's contributions." In addition to reverse transcriptase and the BCR-ABL tyrosine kinase, discoveries that have saved countless lives already, Baltimore has made a third, Nobel-level discovery. This factor is an answer to a riddle. It's not often, I might add, that a scientist has worked at this level—the highest—for three decades. Baltimore was looking for factors that control the machinery of antibody-formation and turned up a transcription factor that was key not only for humoral, but also cellular immunity.[15]

EXPLOSION IN A MUNITIONS FACTORY

NF-kappa B is the answer to a riddle that has puzzled immunologists for half a century: the question of how bacterial endotoxins work and how cortisone cools inflammation. Lewis Thomas described the problem:

> When injected into the bloodstream, endotoxin conveys propaganda, announcing that typhoid bacilli (or other related bacteria) are on the scene and a number of defense mechanisms are automatically switched on, all at once ... including fever, malaise, hemorrhage, shock, coma and death. It is something like an explosion in a munitions factory.[16]

Thomas first encountered the effects of endotoxin in the 1930s when he was a student at the Harvard Medical School. In the depths of the Massachusetts General Hospital he came across a pickled specimen in the pathology department which, "like King Charles's head," would come to haunt his investigative career for decades to come. By accident he had knocked over a sealed glass jar containing the kidneys of a woman who had died late in pregnancy. Returning the jug to its place, he noted that both organs were symmetrically scarred by the deep, black, telltale marks of "bilateral renal cortical necrosis," a rare but fatal complication of infection in late-term pregnancy. Thomas remembered having seen photos of something like those two pock-marked kidneys before. They illustrated lesions that could be produced in rabbits by two appropriately spaced intravenous injections of endo-toxin; the process was called the generalized Shwartzman phenome-non. At the time, Thomas was sure that the phenomenon was "a model for all sorts of events that might be occurring in infectious diseases, then commonplace and insoluble problems for medicine." Finding out what caused the Shwartzman phenomenon was to occupy Thomas's attention—and pages of his bibliography—over the course ·of his career. At a Harvard Medical School alumni day in 1962, Thomas con-fessed exasperation:

> I will not undertake to solve the Shwartzman reaction, even though this problem, like King Charles' head, has been stuck in my mind for over 15 years.... These are, in my view, inviolable secrets of nature, designed by Providence to keep experimental pathologists contentedly engaged for whole lifetimes, and not to be solved.[17]

For three decades (1950–80) Thomas and his many collaborators tried to understand how bacterial endotoxins caused the generalized Shwartzman phenomenon—in fact how endotoxin caused any sort of tissue injury. First, they found out that if one removed white cells from the equation—by various means such as nitrogen mustard—kidney injury was prevented. Next, they found that if they prevented blood from clotting by means of heparin, one could prevent small blood ves-sels from becoming plugged with fibrin, platelets, and white cells: again the kidneys were spared. Those experiments suggested that a combination of humoral and cellular factors made by the host caused the injury. After the tissue had been prepared by the first shot of endo-toxin, a second burst caused fibrin and platelets to plug small blood vessels. White blood cells became trapped in the plugs and discharged proteases which chewed up the matrix of connective tissue.[18] They also tried to understand how cortisone inhibited experimental tissue injury and inflammation, but, alas, the results were less than conclusive.[19] Thomas confessed that he had collected a "good many facts," but had gained little real understanding; the action of cortisone remained a puzzle.

> In a sense, endotoxin shock might be interpreted as a state of ante-mortem autolysis, and cortisone, in doses sufficient to protect the rabbits against shock, blocks the effect of endotoxin. . . . It is now our feeling that there may exist, after all, a single central mechanism which governs all the confusion involved in tissue damage and acts at the core of the inflammatory reaction as a sort of prime mover.[20]

The prime mover turns out to have been Baltimore's NF-kappa B. Leaning heavily on Baltimore's work, we and others have learned that NF-kappa B is the fuse to that "explosion in a munitions factory" and that cortisone works by damping the fuse. Endotoxin exploits genes largely controlled by NF-kappa B to make blood vessel walls, and white cells send out inflammatory signals (cytokines) to make cells stick to each other and the walls of venules, to clot blood and then make it unclottable, to make capillaries leaky, and to fire the whole reaction by inducing fever. Cortisone, in turn, reverses the process by forcing cells to make an inhibitor of NF-kappa B.[21] If you scramble the genes for NF-kappa B, cortisone doesn't work anymore.[22] Right and bright and first with NF-kappa B—as with reverse transcriptase and BCR-ABL—Baltimore has been as much ahead of the curve on inflammation as on AIDS and leukemia.

THE BALTIMORE CASE

Baltimore worked out NF-kappa B while he was enmeshed in a well-publicized, decade-long assault on his integrity by a strange alliance between congressional pettifoggers and censorious Naderites. Following the lead of two professional science vigilantes, Ned Feder and Walter W. Stewart, critics from the Right and the Left of the political spectrum joined the posse. The issue boiled down to whether a coworker of Baltimore's, Dr. Thereza Imanishi-Kari, had kept proper records of experiments on the molecular basis of humoral immunity. The facts became entangled with old-fashioned Nobel envy (what the Germans call Schadenfreude) and soon acquired racist overtones: Imanishi-Kari's Japanese/Brazilian origins were cited gratuitously in the press. Science envy may well have been a factor in calls for Baltimore's resignation from the presidency of Rockefeller University. Barbara Ehrenreich, whose scientific credentials include "a somewhat irrelevant Ph.D. in biology" (from Rockefeller University) wrote in *Time*: "Baltimore should issue a fuller apology, accounting for his alleged cover-up of the initial fraud, and then consider stepping down as president of the university."[23, 24] When one of his supporters had the temerity to rank Baltimore's contribution to our understanding of sepsis (NF-kappa B, again) with that of Dr. Oliver Wendell Holmes (the cause of childbed fever), a snippy reviewer in *The Nation* asked,

> Would you, for example, buy David Baltimore as a meliorist hero? . . . Proffering David Baltimore as a descendant of Oliver Wendell Holmes requires a censor's hidden hard-on for hypocrisy. Holmes, remember, dared to criticize the sanitary habits of his own profession. Baltimore, conversely, has treated criticism of his work as treason.[25]

Hypocrisy or not, calmer heads prevailed, Baltimore and Imanishi-Kari were exonerated, the bloodhounds reverted to their cage, and the science continued.[26]

THE EHRLICH CASE

Paul Ehrlich (1854–1915) went through a similar experience after he had found his magic bullet. Like Baltimore, Ehrlich made three breathtaking discoveries and opened up three fields of science. He was the first to classify cells and microbes according to their affinity for dyes (cytochemistry); he gave weight and number to toxins and anti-toxins (immunochemistry) and he developed Salvarsan as a cure for syphilis (chemotherapy).[27] Like Baltimore, he won a Nobel Prize which made him a target for the less gifted and—again like Baltimore—he was forced to defend his science against vigilantes of the Right and Left. Ehrlich won his prize in 1908 for work on antibodies, the mediators of humoral, or acquired, immunity. He shared the prize with Élie Metchnikoff, who discovered that phagocytosis was the basis of cellular, or innate, immunity. Ehrlich had elaborated a prescient model of how toxins interlock with their antitoxins, the side-chain theory of humoral immunity. He concluded that the reactants could be studied according to the laws of chemistry and physics, a reductive notion that did not endear him to the nationalist philosophers of *Geist* (spirit). His Nobel Prize citation includes a list of questions that remained unanswered at the time: "Why are antibodies only built up against some substances and not against all substances which are foreign to the organism? Where are the antibodies formed? By what process are they formed? What is the nature and constitution of these antibodies? How do they react on the microorganisms and their poisons?"[28] Ironically, the Imanishi-Kari paper at the center of the "Baltimore case" was designed to answer the third of these questions. Ehrlich would have been pleased that it was based on the ultimate of reductive strategies, transgenic animals. Its conclusion, that a transgene can influence a host gene, remains unchallenged today.

Two years after the notoriety of his Nobel Prize, and within a year of Salvarsan's use in the clinic, attacks on Paul Ehrlich ended in calumny and the courts. The German vigilante pack was led by a nationalist physician, Dr. Richard Dreuw of Berlin, and a zealous Frankfurt pamphleteer, Karl Wassman, a "strange-looking man dressed in a dark monk's habit, with a rope around his waist . . . who believed in curing all diseases by Nature alone."[29,30] Wassman's pamphlet *Die Warheit* (The Truth) accused Ehrlich and his Japanese coworker Sahatschiro Hata (1873–1938) of concocting a dangerous, unreliable drug (called 606 at that time) and the Frankfurt hospital of shoddy record keeping. Schadenfreude and racism became an integral part of the story: "Die fachliche Kritik an dem Heilmittel wird mit antisemitischen Angriffen auf Ehrlichs Person verbunden." (Technical critique of the drug went hand-in-hand with antisemitic attacks on Ehrlich himself.)[31] Lutheran clerics argued that the wages of sin was syphilis and that Ehrlich et al. were disrupting the natural order of crime and punishment. More familiar notes sounded from the Left. Ehrlich had signed over manufacture of drug to Hoechst, which charged ten marks—sixty or so dollars—for a course of Salvarsan. The critics complained that Ehrlich was getting rich, that Hoechst was profiting from basic research funded by the government, and that clinical trials of Salvarsan had been carrried out on the prostitutes of Frankfurt without their consent. Things came to a head in a lengthy and drawn-out libel suit brought by the hospital against Wassman on behalf of Ehrlich and Hata. The proceedings turned into a circus as witnesses for the defense were corralled from the red-light district and shadier areas of town. In the end, however, Wassman lost, was sent to prison, and Ehrlich was exonerated. But soon World War I supervened and the Guns of August drowned out the uproar in Frankfurt. Wassman was pardoned, changed the name of his pamphlet to *Die Liebe* (Love), and never mentioned 6o6 again.[32] Salvarsan went on to set the gold standard for the treatment of syphilis until 1937.

Paul Ehrlich, who had been ahead of the curve of his day, identified his opponents at the trial as a pack of "naturopaths, anti-injectionists, quacks, anti-vivisectionists and antisemites."[33] As David Baltimore can testify, they're at it again today. Never mind. I daresay that Gleevec will be to future cancer therapy what Salvarsan was to penicillin: a clue that we're on the right track, at last.

April 3, 2001
Rats, Lice, and History

THIS WEEK, news of typhus came from the clinic, the lab, and the genome network. In New Mexico, epidemic typhus, borne by lice that carry *Rickettsia prowazekii*, was responsible for meningitis.[1] In Texas, endemic (murine) typhus, borne by rat fleas that carry *Rickettsia typhi*, was the rage.[2] Back-to-back bulletins from the lab, published in the March *Infection and Immunity*, reported that *rickettsiae* cause typhus by forcing our blood vessels to send molecular signals that make them targets of choice for our own, self-destructive lymphocytes.[3] Meanwhile, the TIGR/Celera genome brigade finished the complete genome of *C. crescentus*, a missing link in the story of typhus.[4] The genome of *C. crescentus* suggests that our mitochondria began as proteobacteria very much like those of *R. prowazekii*: our ancestors tamed typhus to trap energy from the sun. Untamed, the germs of typhus have written history, as Hans Zinsser, the Harvard bacteriologist taught a generation in *Rats, Lice and History*.[5] Typhus stopped the Turks at the Carpathians, turned Napoleon's Grand Armée back from Moscow, and on or about March 31, 1954, killed Anne Frank in Bergen-Belsen, only two weeks before the British army came in to stamp out the lice.[6]

DEEP IN THE HEART OF TEXAS

Although sporadic outbreaks of endemic, murine typhus have been reported for decades along the Gulf coast of Texas, the disease is now progressing statewide.[7] Health officials elsewhere in the nation—in New England, the Midwest, and California—worry about tick-borne diseases such as Lyme disease, ehrlichioses, and Rocky Mountain spotted fever. Not in Texas, though, where 211 cases of murine typhus were reported from 1995 to 1998 as against 18 cases of Rocky Mountain spotted fever and 6 of ehrlichiosis.[8] Murine typhus is carried from rats to humans by the rat flea, *Xenopsylla cheopis*. When the flea bites, it drops its fecal load of microbes on the skin, which the itchy victim rubs into the puncture. The majority of patients with murine typhus in Texas are Hispanic youngsters, the "classical" triad of fever, headache, and rash was found in only half the cases, while gastrointestinal symptoms were present in almost 80 percent; many showed signs of multiple organ system involvement including heart, liver, kidney, blood, and central nervous system. But the Texas doctors were on the lookout for this disease; severe complications were rare and prevented by early treatment with tetracyclines and chloramphenicol.[9] Another threat was reported from New Mexico where a severe case of meningitis was first diagnosed as murine typhus until an ingenious application of the polymerase chain reaction identified the culprit as *R. prowazekii*. Epidemic typhus is transmitted from human to human—and in rare cases from flying squirrels to humans—by the louse *Pediculus humanus humanus*, a strictly human parasite that lives and thrives in clothing.[10]

Lice require dirty humans, bad weather, and crowding—as in tents and barracks. That's why typhus is the stuff of war and tragedy. During World Wars I and II, typhus spread through North Africa, the Pacific Islands, and Europe, where it was the second leading cause of death in the German concentration camps. U.S. forces were protected by a vaccine based on one developed by Hans Zinsser and applied on a large scale. Although epidemic typhus declined at the end of World War II with the advent of DDT, *R. prowazekii* is making a comeback.[11]

The largest recent outbreak since World War II was in Burundi in the mid-nineties, where modern molecular techniques were used to show that a single outbreak of "jail fever" in Burundi sparked an extensive epidemic of louse-borne typhus in the refugee camps of Rwanda, Burundi, and Zaire—countries racked by ongoing civil war and genocide.[12] There was also a brisk outbreak in Russia in 1997. In Europe in the past and in Africa today, persons who have "recovered" from epidemic typhus in their youth suffer relapses of the disease (Brill-Zinsser disease) to become a reservoir of new, louse-borne epidemics.[13] When the body politic breaks down, hygiene loses and lice prevail.

DEEP IN THE HEART OF VESSELS

We know a good bit these days about rickettsia and how they produce disease. The bacteria cannot survive outside living cells, they are "obligate endosymbionts." Indeed, the bugs are happiest in the intestinal cells of lice where they live as harmless parasites. Not so in humans: when the typhus germs enter our system it spells trouble for host and parasite alike. In the absence of antibody, the bacteria induce the lining cells of our blood vessels, endothelial cells, to form digestive pockets called phagosomes. Refreshed by nutrients of the host, the microbes secrete a membrane-lytic substance—a "hemolysin" which ruptures the phagosome ro release rickettsiae into the cell sap where they are free to divide and multiply. In response to this blight, the cells of the host send out signals of distress and inflammation: prostaglandins, chemoattractants, chemokines of several sorts. In the end, the host's own immune cells inflict most of the damage; macrophages and lymphocytes attack the infected blood vessel cells which display on their surface antigens derived from the bacterium.[14] This assault kills infected cells, but also presents fragments of the microbe to the host's antigen-processing machinery. Once antibodies kick in, recovery begins. Pathologist David Walker of University of Texas Medical Branch at Galveston has been studying these events for decades and this week announced that he had figured out which of the many inflammatory cells was critical to recovery from experimental rickettsial infection. The culprit was the CD8+ cytotoxic lymphocyte, previously known for its role in immune surveillance against tumors and viruses. Moreover, he identified which protein in the cytotoxic cells does the job: it's another membrane-lytic protein called perforin. Mice in which the perforin gene was deleted were a hundred times more susceptible to *rickettsia* than the wild type.[15] All that furor in the blood over a microbe with fewer than eight hundred genes.

THE MISSING LINK IN MITOCHONDRIA

Since 1970 most biologists have accepted Lynn Margulis's surmise that our mitochondria are residues of proteobacteria that took up residence in some distant unicellular ancestor of ours.[16] Surmise turned to probability in late 1998 when a Swedish team published the complete genome of *R. prowazekii*.[17] The 1.1 million base pairs of the typhus DNA genome contained 834 protein-coding genes, among which were all the genes for the Krebs cycle and the respiratory-chain complex of human mitochondria. Not only that, but phylogenetic analyses (ribosomal DNA) showed that *R. prowazekii* was "more closely related to mitochondria than to any other microbe studied so far." The Swedes guessed that there must have been an early evolutionary event when the bug's genes were off-loaded from some such typhus-like bug to the nucleus of an animal cell. What would such a missing link look like?

This week, Craig Venter, Claire Frasier, and their teams published the results of a collaboration with the Susan Sontag of molecular biology, Lucy Shapiro of Stanford University.[18] Shapiro has spent decades working out the intricate cell-cycle machinery of *C. crescentus*, a free-living member of the alpha group proteobacteria, closely related to such human pathogens as *B. abortus*—and to *R. prowazekii*.[19] *C. crescentus* turns out to have four million base pairs encoding 3,767 genes. Since the microbe coordinates its cell division cycle and such cellular gymnastics as chemotaxis, flagellar movement, and the like—

tricks unheard of in symbionts like the typhus bug—it was not surprising that the *C. crescentus* genome encoded a significantly higher number of signaling proteins than any bacterial genome sequenced thus far. The genome hunters announced that they had shown the "close relationship between *C. crescentus* and the nonfree-living endosymbiont, *Rickettsia prowazekii*, the only other sequenced-proteobacterium." They had found a missing link between bacteria and mitochondria, an "off-loading" precursor of the *rickettsiae*. Modestly, they predicted that "those genes critical to cell cycle progression in *rickettsia* are less likely to have been lost during its reductive evolution."

REDUCTIVE EVOLUTION AT BERGEN-BELSEN

"Reductive evolution" applies to the history of typhus in central Europe. Even before *rickettsiae* were identified, it was appreciated that epidemiology and public health might be better weapons against typhus than drugs or antisepsis. In 1848, a young Rudolph Virchow (1821–1903), the future father of cellular pathology and future leader of the Social Democrats in the Berlin parliament, was sent to study an intractable typhus epidemic in Upper Silesia, at the eastern border of Germany and Poland. In this backward enclave, feudal landlords of large estates ruled a peasantry in dirty huts. Virchow reported to Berlin that "the Upper Silesian in general does not wash himself at all, but leaves it to celestial providence to free his body occasionally by a heavy shower of rain from the crusts of dirt accumulated on it. Vermin of all kinds, especially lice, are permanent guests on his body." Virchow concluded that what the region needed was not more doctors; it required social medicine for a social disease. Typhus could only be eradicated, Virchow argued, by means of political reform: full employment, higher wages, the establishment of agricultural cooperatives, universal education, and the disestablishment of the Catholic Church.[20] And slowly but surely, from Bismarck's universal health insurance to the Weimar Republic's secular reforms, typhus abated. Alas, less than one hundred years later, reductive evolution had "off-loaded" the genes for social democracy in Germany and typhus returned. Its most poignant episode was played out between March 31 and April 15, 1945, at Bergen-Belsen. Anne Frank was one of eight Dutch Jews who had been in hiding for two years and thirty days when they were discovered and arrested by the Nazis and deported from Amsterdam to Auschwitz-Birkenau. Records have it that Anne's mother died January 6, 1945, at Auschwitz-Birkenau (Netherlands Red Cross, dossier 117265). As the Russian armies moved west, the Germans forced thousands of men, women, and children across war-torn Europe to Bergen-Belsen, a desolate concentration camp north of Hannover.[21] Among them were Anne and her elder sister, Margot, who died of typhus sometime around March 31, 1945, at Bergen-Belsen concentration camp (Netherlands Red Cross, dossiers 117266 and 117267).[22] Approximately fifty thousand of their fellow prisoners died, most of disease, most of typhus. The last woman to see the sisters alive described the hut in which they were quartered:[23]

> The dead ones were always carried out and put in front of the barrack. When you were permitted out in the morning to go the latrine, you had to pass them. That was just as horrible as making your way to the latrine, because almost everyone had typhus. In front of the barrack there stood also a kind of wheelbarrow, into which one could also drop your excrement. . . . Probably on one of those trips to the latrine or barrow I passed the corpses of the Frank sisters, one or both, I know not. I assumed at the time that the corpses of the Frank girls had been stacked in front of the barracks. And those stacks of corpses were cleared away. There was a large pit dug in the camp and they must have been thrown into it.[24] (Es wurde eine große Grube gegraben, da wurden sie hineingeschmissen, so kann man es wohl sagen.)

On April 15, 1945, Bergen-Belsen was liberated by the Allied 21st Army Group, a combined British-Canadian unit. At the time of liberation, the camp had been without food or water for three to five days.

Although the camp commandant, Josef Kramer, protested that there was no way to pipe water into the camp, the Allied 21st quickly constructed a makeshift piping system from a nearby river, to supplement army water carts. Despite the best efforts of Allied relief workers, more than ten thousand seriously ill inmates died after liberation, again mainly from typhus. Sixty thousand prisoners were liberated by the Allies. The Bergen-Belsen staff, captured en masse by the 21st on the day of liberation, were tried in 1945 by a British military tribunal in nearby Lüneburg. Among those tried were *Kommandant* Josef Kramer and a twenty-two-year-old female S.S. Guard, Irma Grese, who was accused by camp inmates of shooting prisoners and beating them with a homemade whip.[25] The diaries of the liberators tell the story of the Belsen typhus epidemic and the reductive evolution of social democracy to national socialism.[26] From the diary of volunteer medical student Dr. Michael Hargrave, April–May 1945:

So we reported back to the Office and we were told to try Hut 224 Laager 1 (Women). . . . We went into the hut and were almost knocked back by the smell, but we went into one of the two main rooms. The sight that met us was shocking—there were no beds whatsoever and in this one room there were about 200 people lying on the floor. In some cases they wore a few battered rags and in some cases they wore no clothes at all. . . . Their hair, hands, faces and feet were all covered in a mixture of dry faeces and dirt. Here and there a dead person could be seen lying between two living ones, who took no notice of her at all and just went on eating, coughing or just lying, and these were all women whose ages varied from 15–30.[27]

From the diary of Lt. Col. Mervin Willett Gonin DSO:

I can give no adequate description of the Horror Camp in which my men and myself were to spend the next month of our lives. It was just a barren wilderness, as bare and devoid of vegetation as a chicken run. Corpses lay everywhere, some in huge piles, sometimes they lay singly or in pairs where they had fallen. Those who died of disease usually died in the huts. Piles of corpses, naked and obscene, with a woman too weak to stand proping herself against them as she cooked the food we had given her over an open fire; men and women crouching down just anywhere in the open relieving themselves of the dysentery which was scouring their bowels, a woman standing stark naked washing herself with some issue soap in water from a tank in which the remains of a child floated.[28]

From central Europe to Burundi, perhaps even in the heart of Texas, typhus is what Rudolph Virchow believed it to be: a social disease.

March 21, 2001
Hitler's Gift and
the Price of AIDS

THIS WEEK Big Pharma was in trouble on two fronts. First, the Dow Jones industrial average had its biggest weekly drop in eleven years and shares of large pharmaceutical firms plummeted. Worse yet, they were forced to rethink their price structure. Faced with an unstemmed plague of AIDS in Africa, and with activist pressure that spread from Durban to New Haven, Merck and Bristol-Myers Squibb cut the cost of anti-AIDS drugs for sub-Saharan Africa, where more than 25 of the 36 million people infected with HIV worldwide live.[1] It was years too late; not since Big Tobacco was in the dock has industry looked more craven. On the other hand there was some good news for Big Pharma. Two new pharmacoeconomic studies showed that aggressive drug treatment of AIDS works and that it costs less than simple supportive care, at least in the United States.[2] Those ups and downs of the market-place cannot tarnish the clear achievements of rational science. We owe much of antiretroviral drug design, of the rogue molecules one can splice into DNA, to a generation of refugee scientists that Jean Medawar and David Pyke have called "Hitler's Gift."[3] One of those gifts was Erwin Chargaff (1905–93), the Columbia biochemist who fled Nazi Austria to discover from the DNA of creatures great and small that the sum of adenine and thymine equals the sum of guanine and cytosine.[4]

COCKTAIL HOUR

Modern therapy for AIDS involves three classes of drugs—two directed at the reverse transcriptase of the virus, the other at the protease required for its procreation. The optimum cocktail uses one of each of these :

- The first of these that FDA approved was AZT (zidovudine) in 1987, a reverse transcriptase inhibitor of the nucleoside analogue class. Others include ddI (didanosine), ddC (zalcitabine), d4T (stavudine), 3TC (lamivudine), and Ziagen (abacavir). In 1997, FDA approved Combivir, a mixture of AZT and 3TC which allows patients to reduce the number of pills needed.
- Viramune (nevirapine), the first reverse transcriptase non-nucleoside analogue inhibitor, was approved in 1996, followed by Rescriptor (delavirdine) and Sustiva (indinavir).
- Protease inhibitors have only been on the market about four years. Invirase (saquinavir) was followed by Norvir (ritonavir), Crixivan (indinavir), Viracept (nelfinavir), and Agenerase (amprenivir).[5]

In the United States, aggressive treatment with the triple-drug cocktail renders the virus undetectable in most patients by two to three months. But in undeveloped parts of the world, or in underserved parts of our country, drugs are neither available nor affordable.[6] Following several weeks of active, visible protests on the steps of courtrooms in South Africa and on the campus in New Haven, Big Pharma caved in.[7] Bristol-Myers Squibb announced this week that it would not prevent generic-drug manufacturers from selling low-cost versions of d4T (stavudine)—which it calls Zerit—in Africa. Bristol-Myers also said it would cut the price of Zerit and ddI (didanosine), or Videx, to a combined price of a dollar a day in Africa. In the United States one day's dose of the two goes for eighteen dollars.

Yale shares the rights to d4T with Bristol-Myers thanks to the pioneering work of William H. Prusoff, a star in perhaps the most distinguished department of pharmacology in the world. Prusoff's work on base analogues has been a source of new compounds and newer ideas since the '6os.[8] Despite sharing the patent rights to d4T, Prusoff joined AIDS activists in urging Yale and Bristol-Myers to lower its prices in Africa.[9] Yale law students threatened, university officials waffled, TV spread the news, and spring was in the air. Yale chose altruism over avarice: it went along with Bristol-Myers's decision to cut the tab.

"This is not about profits and patents," said newly altruistic John L. McGoldrick, executive vice president at Bristol-Myers. "It's about poverty and a devastating disease. We seek no profits on AIDS drugs in Africa, and we will not let our patents be an obstacle."[10]

"This is groundbreaking," said Kate Kraus, a member of Act-Up Philadelphia, a group that has led protests around the world against the big drugmakers. "This is the first time that a U.S. drug company has acknowledged that generic drugs are the key to saving lives."[11]

Bristol-Myers Squibb's announcement came a few days after Merck had announced that it would sell Crixivan (indinavir) and Sustiva (indinavir) to undeveloped countries for about a tenth of the U.S. price. Crixivan costs $6,016 per patient each year in the United States, but will go for $600 per patient per year; Sustiva will go for $500 instead of $4,730 in the United States.[12] The Treatment Action Campaign, a South African AIDS activist group, wasn't all that impressed by the altruistic motives of industry, and argued that only their intense pressure had forced Big Pharma to capitulate: "This victory has come about as a result of the global effort by HIV/AIDS activists."[13] Perhaps, but the companies weren't giving up all that much. It turns out that no more than 10 percent of Big Pharma income is from the developing world, and only 1.6 percent from Africa.[14] It's taken them a long time to come around, but by hook or by crook, by altruism or activism, the drugs are spilling out.

HOW MUCH DOES A YEAR OF LIFE COST?

Meanwhile, in the United States, two new studies in this week's NEJM show us that effective drugs are cheap at any price. Bozzette et al. used applied sociology of the sort favored by the Rand people.[15] They interviewed a random sample of 2,864 patients receiving care for HIV infection from 1996 to early 1999. At baseline, the mean expenditure was $1,792 per patient per month, but by 1997 it had already declined to $1,359 for survivors. They found that increases in drug expenses with aggressive treatment were far less than the reductions in hospital costs. The good news was that the more agressive the drug treatment was, *the less the overall cost.* Overall, the estimated annual expenditure declined from $20,300 per patient in 1996 to $18,300 in 1998. Unfortunately, social factors played a role: drug costs were lowest, but hospital costs highest, among "underserved groups, including blacks, women, and patients without private insurance." But the happy conclusion—for those who can afford it—was that: "the total cost of care for HIV infection has declined since the introduction of highly active antiretroviral therapy."

A second study in the NEJM was from the field of theoretical pharmacoeconomics.[16] Freedberg et al. estimated that as of June 2000, more than three hundred thousand people were living with AIDS. Each year, there are about forty thousand people newly infected with HIV and sixteen thousand deaths from AIDS. Dramatic declines in the incidence and deaths from AIDS began in the United States in 1996, when the triple-drug cocktail entered the clinic. What happened to costs? Examining truckloads of massaged data, Freedberg's Massachusetts General Hospital group estimated that life expectancy "adjusted for the quality of life jumped from 1.53 to 2.91 years, while per-person lifetime costs increased from $45,460 to $77,300 with the cocktail as compared to supportive treatment." The increased cost per "quality-adjusted year of life gained" was $23,000. The authors concluded that "Treatment of HIV infection with a combination of three antiretroviral drugs is a cost-effective use of resources."

I'm ready to go along with that. At $23,000 per person per year, it's worth it. It'll be cheaper by far when we discover a real cure. As Lewis Thomas wrote shortly after Apollo landed:

If I were a policy-maker, interested in saving money for health care over the long haul, I would regard it as an act of high prudence to give high priority to a lot more research in biologic science. This is the only way to get the full mileage that biology owes to the science of medicine, even though it seems, as used to be said when the phrase still had some meaning, like asking for the moon.[17]

HITLER'S GIFT

Speaking of full mileage, I recalled a symposium I attended with Bill Prusoff in 1967.[18] It had been convened in London by Hermann Blaschko, a genial Oxford pharmacologist who would be celebrated along with his fellow refugees from Nazidom in Medawar and Pyke's marvellous elegy, "Hitler's Gift." Prusoff had begun his symposium lecture with the famous base-pairing diagram of Chargaff; he went on to speak of the overall impact of Chargaff's work of the '60s, work that prompted many of us to pursue studies of the template activity of neat and substituted DNA.[19, 20] The DNA chemists gathered there were surprised that Chargaff had not won all the glittering prizes for his work. Blaschko thought that Chargaff ranked with the once and future Nobel Prize winning emigres who had escaped Hitler for the labs of Britain and the United States: Otto Loewi, Ernest Chain, Konrad Bloch, Hans Krebs, Max Perutz, Severo Ochoa, Fritz Lipmann, Salvador Luria, and Max Delbrück. Opinion at the time was, however, much influenced by James Watson, who seemed to have had the final word in *The Double Helix*:[21]

The moment was thus appropriate to think seriously about some curious regularities in DNA chemistry first observed at Columbia by the Austrian-born biochemist Erwin Chargaff. . . . In all their DNA preparations the number of adenine (A) molecules was very similar to the number of thymine (T) molecules, while the number of guanine (G) molecules was very close to the number of cytosine (C) molecules. Moreover, the proportion of adenine and thymine varied with their biological origin. No explanation for his striking results was offered by Chargaff, though he obviously thought they were significant.

Not true. Chargaff had indeed figured out that the variations in A-T and G-C content might dictate genetic information:

Desoxypentose nucleic acids from different species differ in their chemical composition, as I have shown before; and I think there will be no objection to the statement that, as far as chemical possibilities go, they could very well serve as one of the agents, or possibly as the agent, concerned with the transmission of inherited properties.[22]

Chargaff had turned the transforming principle of Oswald Avery, Colin Macleod, and Maclyn McCarty into the new chemistry of DNA.[23] In painstaking work, he had spelled out the ratios of bases in the DNA of species ranging from bacteria to mammals. Mac McCarty agreed. In his 1986 memoir he gave Erwin Chargaff the credit he deserved: "This discovery of base pairing adenine-thymine and of guanine-cytosine was of prime importance for understanding the structure of DNA, and it proved to be a decisive factor in the formulation of the Watson-Crick model." Nowadays we also know that if Chargaff hadn't worked out the chemistry of base pairing, DNA chemists like Prusoff wouldn't have been able to show how and where those analogues of thymidine inserted.[24] And by the time d4T was born, Prusoff harnessed chemists from Bristol-Myers to launch this new analogue based on the principle that rupture of Chargaff's base pairing would inhibit retroviral replication.[25] It was another gift of Hitler's grotesque regime.

March 7, 2001

Foot-and-Mouth Disease:
Flames over England

THIS WEEK virulence made news the world over. In Afghanistan, Mullah Mohammad Omar, leader of the Taliban militia, ordered destruction of all statues in the country and began shelling two of the world's tallest standing statues of Buddha because "God is one God and these idols have been gods of the infidels."[1] In England, heeding warnings that "there's a panic" in the country, Tony Blair suspended horse racing, canceled a rugby match between Ireland and Wales, and ordered the mass destruction of pigs, cattle, and sheep.[2] In Belgium, police turned fire hoses on farmers who had stormed barricades around the Brussels headquarters of the EU where agricultural officials were in emergency session.[3] The action in Kabul was a virulent aspect of the religious strife that has plunged Afghanistan into the state of nature; the actions in London and Brussels were due to a virus that persuades its host (be it cloven-hoofed or human) to make a papain-like enzyme that picks its cells apart from within.

IN HERTFORD, HEREFORD, AND HAMPSHIRE

Flames filled the night sky over England as thousands of animals were incinerated after Britain was struck by an epidemic of hoof-and-mouth disease.[4] As if a decade of mad cow disease weren't enough (see 11/27/00), along came foot-and-mouth disease, properly dubbed "one of the remaining great plagues."[5]

The wasting disease, characterized by blisters in the mouth and feet of cloven-hoofed animals, has a case fatality rate approaching 50 percent in young animals and is caused by a picornavirus closely related to polio or rhinoviruses. Although it spreads but infrequently to humans, it is highly contagious for sheep, goats, cattle, and pigs.[6] The disease was first detected in twenty-eight pigs at an Essex abattoir, then sheep came down with it in Devon, then lambs in Lancashire, and cattle and sheep in Wiltshire and Hereford. Hereford? Well, perhaps in Hertford, Hereford, and Hampshire, Hurricanes Hardly Happen, but foot-and-mouth disease hit the fan. By March 7, according to Britain's chief veterinary officer, Jim Scudamore, the disease had spread to over eighty-four farms and abattoirs all over the island. The total number of animals slaughtered or due to be destroyed in Britain is over seventy-five thousand, with an additional 210 farms around the country under restrictions. It has crossed the borders to Scotland and Northern Ireland.[7]

The outbreak of foot-and-mouth disease was Britain's first since 1967, when 442,000 animals were slaughtered in five months, and this time the authorities have responded with draconian measures. Since the virus is airborne and can also be carried on shoes or clothing, government veterinary officers have set up five-mile exclusion zones for people and vehicles at all farms or abattoirs where infected animals are found. The St. Patrick's Day parade has been canceled in Dublin, and Dolly the cloned sheep has been quarantined in Edinburgh.[8] In England, tourists walk through disinfectants, trucks are sprayed in the chunnel, and officials have closed, among others:

- Public footpaths and rights of way in the countryside
- Wales v. Ireland six-nations rugby match in Cardiff
- Crufts, Britain's biggest dog show (Birmingham from March 8 to 11)
- The national forests, national parks, and the deer sanctuary at Epping Forest, Essex

- All zoos, including Whipsnade Wild Animal Park in Bedfordshire
- Hampton Court in London and Magdalen College, Oxford
- All hunting by harriers, beagles, foxhounds, and staghounds
- Fishing in Herefordshire, Wales, Cheshire, Lancashire, Cumbria, and Yorkshire[9]

After a lengthy meeting in Downing Street, Blair's cabinet prohibited movement of all livestock within the country until March 16. It also approved fines of up to £5,000 ($7,200) to ensure observation of trespass limits for people and vehicles. In consequence of these measures, and since the government has urged people to keep out of the countryside altogether, farmers have no markets, hikers are homebound, and children are crawling the walls. Sad to say, the epidemic need not have hit Britain at all. Mass annual vaccination against foot-and-mouth disease, previously applied by the eight-member states in the European Community, was progressively phased out during 1990–91. Because of the vaccine's cost and doubts over its efficacy, the other four member states (the United Kingdom, Denmark, the Republic of Ireland, and Greece) either never have vaccinated or ceased to do so several years before 1990. Right after the EC stopped vaccinating, Dr. Alex Donaldson, head of the UN's world reference laboratory for the virus, had warned that "cessation of vaccination will result in a higher proportion of fully susceptible cattle and in the event of outbreaks will increase the likelihood of the rapid dissemination of virus and increase the risk that the infection will enter Great Britain."[10] He was right: the lambs were ready for slaughter.

I'VE NEVER MET A RUDER PEST

Humans are also susceptible to foot-and-mouth disease; the virus has been isolated and typed in more than forty human cases. This virus was in fact the first animal virus ever described (in 1897, by Friederich Löffler, who four years earlier had discovered the diphtheria bacillus). Animal-borne foot-and-mouth disease in humans is most often confused with infections caused by the Coxsackie A group of enteroviruses (referred to as "hand, foot, and mouth disease"), often by herpes simplex, and sometimes by vesicular stomatitis virus. It's a modest, febrile disease with a good prognosis. Criteria for establishing a diagnosis of foot-and-mouth disease in man are the same as in animals: isolation of the virus from the patient and/or identification of specific antibodies after infection. Proven cases of foot-and-mouth disease in man have been confirmed in several countries in Europe, Africa, and South America. Serotype O (the type now floating around England) is the most common, followed by type C and rarely A. The incubation period in man, although somewhat variable, is between two days and six days.[11] Humans are less susceptible than livestock, because while the virus attaches to the same family of surface receptors, integrens alpha (v) beta (3), on bovine and human cells, it interacts less avidly with the human beta subunit.[12]

The foot-and-mouth disease virus is a member of the picornavirus family of positive-strand RNA viruses; these include poliovirus, hepatitis A virus, rhinovirus, and encephalomyocarditis virus. Its single-stranded, positive sense RNA genome expresses a single protein of around two thousand amino acids, designated the polyprotein. This precursor is subsequently cleaved into the mature viral proteins by proteinases coded in the RNA. The properties of the three defined proteolytic activities in picornaviruses are most closely related to papain, although there are subtle structural and biochemical features that distinguish the foot-and-mouth disease virus leader proteinase from other papain-like enzymes.[13] When picornaviruses infect cells, they disrupt protein synthesis by cleaving a translation factor called eIF4G which is responsible for recruiting mRNA to the 40S ribosomal subunit during initiation of protein synthesis. eIF4G is also targeted for cleavage by caspase-3 during apoptosis, but the picornavirus doesn't need caspases to do the job. It does so very nicely by itself.[14] Conclusion: Foot-and-mouth disease virus affects man and beast. This rude pest kills cells by forcing them to make a papain-like enzyme that induces apoptosis. Instructed by the virus, the cell chews itself to death.

THE IMPORTANCE OF THE INFINITELY LITTLE

Europe took prompt steps to prevent cross-channel contamination. German, Dutch, and Belgian authorities ordered animals imported from Britain to be slaughtered as a precaution. European Union veterinary specialists ordered a complete, temporary ban on British livestock, while France announced on its own that it would destroy twenty thousand sheep imported from Britain. German health officials slaughtered fifteen hundred sheep and lambs at a farm near Düsseldorf, after eighty of the sheep were identified as originating from British farms infected with the disease. Since thousands of British sheep and pigs had been imported into Germany and Holland, the governments suspended farm meetings and ordered movement of animals held to a minimum. Germany also confiscated sandwiches, milk containers, and food products carried by English air passengers at Tegel aiport in Berlin, while Belgium banned the transport of all sheep and goats inside the country and barricaded its government offices against its angry farmers.[15] Meanwhile, no extra measures have been taken in the U.S. Presumably due to close monitoring of food products that enter our borders, North America has remained largely free of the disease; the last major outbreak in the United States was in 1929.[16]

The finger of blame quickly pointed at dirty deals by other parties. In the last two years more than sixty countries had outbreaks, among them Israel, Greece, Japan, Brazil, and Uruguay, all of which were previously free of the disease. Alf-Eckbert Fussel of the European Commission's Health Directorate argued that the virulent serotype of the disease found in Britain was not introduced by legal trading "but probably because of smuggling and illegal trade." Last year's outbreak in Japan, the first in more than seventy years, was traced back to diseased straw brought illegally from China, via Russia, while outbreaks in southern Brazil, also last year, led to one in Uruguay, its first in ten years. This week's epidemic in Britain has been traced to serotype O, the "pan-Asia strain," first detected in Taiwan in 1999 and spread within the year to Japan and South Korea.[17]

With panic abroad in the land, jingoism joined racism as Britain's

Ben Gill, president of the National Farmers' Union, suggested that the yellow race was responsible for the British outbreak. "Is it a coincidence that we had classical swine fever in East Anglia last year of an Asian origin, and foot-and-mouth now, also of an Asian origin? It raises questions about freer world trade."[18] The German tabloid *Bild*—a rag that on March 1 split its front page between buxom, unclad *Katie im Kampfanzug* (Katie in battledress) with a story on *Seuchen-Alarm* (short for *Maul und Klauenseuche*, foot-and-mouth disease)—complained that vets were forced to kill seventy-five little lambs in Aachen because of contact with unwashed regions of the world: "Probably a tourist brought it in—from Africa or Asia. Perhaps it arrived in a sandwich. It was thrown away and an animal gobbled it up."[19]

Those sentiments remind me of fears current in the days of the microbe hunters, when all disease came out of Africa, Asia, or the Middle East. Fear of aliens became conflated with fears of the microbes that brought plague, cholera, and typhus from "the Levant" or other places that white men feared to tread, as in Kipling's "The White Man's Burden."

> The ports ye shall not enter,
>   The roads ye shall not tread,
> Go, make them with your living
>   And mark them with your dead.
> Take up the White Man's burden,
>   And reap his old reward—
> The blame of those ye better
>   The hate of those ye guard—
> The cry of hosts ye humour
>   (Ah, slowly,) toward the light:—
> "Why brought ye us from bondage,
>   Our loved Egyptian night?"[20]

That notion was seconded in an unlikely text of Kipling's day. In December 1892 there appeared in *The Bookman* a review of Dr. Arthur

Conan Doyle's recently published *The Adventures of Sherlock Holmes.*
The piece was written by Dr. Joseph Bell, a professor of surgery at the
University of Edinburgh. Since readers of that popular literary magazine knew that the Scot surgeon was the real-life prototype of Sherlock
Holmes—Doyle had been Bell's student and assistant—they also
knew they were reading a review of *The Adventures of Sherlock
Holmes* written by Sherlock Holmes's alter ego. Bell described the
spread of disease in terms that might have pleased farmer Gill or the
editors of *Bild*:

> The importance of the infinitely little is incalculable. Poison a well
> at Mecca with the cholera bacillus, and the holy water which the
> pilgrims carry off in their bottles will infect a continent, and the
> rags of the victims of the plague will terrify every seaport in Chris-
> tendom.[21]

Considering that no single cow, lamb, or pig in Europe need ever
have contracted the hoof-and-mouth disease had vaccination not been
abolished in Western Europe and if it were available cheaply elsewhere in the world, perhaps the seaports of Christendom would not be
terrified today. Not Christendom alone, one might add. Bell would
have appreciated the irony of a ban announced this weekend by China,
Japan, and South Korea on the entry of livestock from Britain into
their seaports: take up the Asian's burden and reap his old reward . . .

February 21, 2001

The Genome Is On-line

AT A WASHINGTON press conference timed to coincide with the 192nd birthday of Charles Darwin, two rival groups of scientists announced that the map of the human genome was now available in print and on-line. Special issues of *Science* and *Nature* published details of how the origin of our species has been spelled out in the letters A and T and G and C. The race for the genome is over. Indeed, after pictures of J. Craig Venter of Celera, Francis Collins and Eric Lander of the International Genome Project wound up on the front pages of newspapers worldwide, the race for the genome could be said to have ended in a photo-opportunity finish. It all strikes me as a replay of the race to discover the AIDS virus, the polio vaccine, or even the North Pole: the magnitude of each discovery dwarfed the acrimony among the discoverers.

CHILLS RUN UP AND DOWN MY SPINE

"I've seen a lot of exciting biology emerge over the past 40 years. But chills still ran down my spine," wrote David Baltimore in *Nature* apropos of the announcement.[1] He was right; but while there was glory enough for all, there was also irony aplenty as the public/private debate erupted over the "Rosetta Stone of Science."[2] A "public" company, Celera, responsible to its many thousands of investors, published its findings in *Science*, a nonprofit journal of the American Association for the Advancement of Science (AAAS).[3] The "public" international consortium, responsible in the United States, France, and Germany to a few political appointees and in the U.K. to a private philanthropy, published its findings in *Nature*, "one of the most commercially astute journals around."[4, 5] Scientists can obtain any full-length article from *Science* for free, but with the exception of its heavily promoted genome issue have to pay for material from *Nature*. Celera's map was based in part on sequence data placed in the public domain by the international consortium; on the other hand, the consortium could not have finished its sequences without machinery developed by Mike Hunkapiller, a coauthor of Venter's and head of Celera's sister company, Applied Biosystems. The British side of the public consortium was funded by the Wellcome Trust, the endowment of which derives from a private pharmaceutical fortune. Eric Lander, the lead author of the consortium's paper, heads the Genome Center of the Whitehead Institute, a gift to MIT, of Edwin C. (Jack) Whitehead, a biotech venture capitalist. The late Jack Whitehead set up the institute and assembled its glittering staff in 1984 over intense objections by much of the MIT faculty to the "commercialization of science." After his earlier offers to endow a Whitehead Institute at either Duke or NYU had been rejected, Jack complained to me at the time that "It's easier to make a hundred million dollars than to give it away." MIT was the winner of that one. Given that the public and private spheres are so intertwined, I tend to agree with Tim Dexter of the Wellcome Trust that the human genome is a "gift to the world" from both sectors, and worth somewhat more than $100 million.[6]

ALADDIN'S LAMP IS MINE

So there they were in Washington, both parties cheek by jowl. Over the course of the last year, Venter had rightly claimed that Celera's "whole genome shot-gun method" had propelled the entire enterprise. Many of us remember that when Francis Collins had assumed leadership of the NIH-led project in 1993 he pleaded the urgency of his plan to finish the genome by 2005, arguing that "Medical advances will be delayed. The U.S. competitive position will suffer. The Japanese are ramping up their efforts."[7] Beaten to the punch, his teammate still maintains that Celera has missed the mark: "All the king's horses and all the king's men could not put the genome together again," quipped Eric Lander in defense of the consortium's cloning strategy. "Instead of a complete map of DNA, Venter wound up with a tossed genome salad."[8]

Venter fought back: "No small amount of this was the politics and psychology of being able to stay with this and stick with it. If there was any way to stop this, it was tried, down to the end of trying to block our paper being published in *Science*. If we weren't resistant and somewhat defiant this never would have gotten done." Venter added that if Hunkapiller had not developed the automated DNA sequencer, "it is unlikely that anybody would be standing here today to discuss what we found on sequencing the human genome."[9] The consortium was by no means unified, nor lacking in the spheres of politics and psychology. Lander was quoted by the *Wall Street Journal* as saying that "Celera never imagined we could be a powerhouse." The *Journal* reported that the Whitehead's "take-no-prisoners attitude" was in evidence when the center secretly started sequencing DNA that had been assigned to other groups in the consortium. In fact, each of the twenty-three human chromosomes had originally been claimed by different labs. "But with the other project leaders still unconvinced of Whitehead's growing capacity and withholding information that would tell it what DNA to study, Dr. Lander's group simply began seizing territory."[10] Venter, an ex-navy corpsman in Vietnam, struck a tougher pose for an English reporter. "Everybody keeps wanting to turn this into a pissing contest of whose is bigger and whose is better, and that's never what it has been about in the first place. . . . Our goal was to get, as I said earlier, the highest quality product we can."[11] Donald Kennedy, the editor of *Science* and a veteran of bitter controversies at Stanford, put the best face possible on the acrimony: "Thus, we can salute what has become, in the end, not a contest but a marriage (perhaps encouraged by shotgun) between public funding and private entrepreneurship."[12]

I've now been able to browse both sets of web sites.[13] As an academic, I'm entitled to full and free use of both data bases and, thanks to a Celera demonstration at NYU, I've looked at the company's for-profit supplemental informatics package. It doesn't hurt Celera that its sites are better designed and easier on the eye. Indeed the Celera system is generally more user-friendly with better links to protein-protein interactions and to searches for protein motifs. Moreover, I can go from protein to gene back to protein again, which is very helpful in looking for what are called "paralogues," sequences in genes and/or proteins that look like known drug targets (e.g., CysLT 4 for leukotrienes). But that's personal preference. What do the published genome sequences tell the community of science? Both groups agree on an astonishing set of findings, none of which were obvious until both sites came on-line, which I'll summarize here:

- The sequences are over 90 percent complete for the euchromatin regions of human chromosomes. In nine months Celera generated a 14.8 billion base pair DNA sequence from 27,271,853 high quality sequences from both ends of plasmid clones derived from the DNA five individuals. Obtained by two techniques—the whole-genome assembly and a regional chromosome assembly—the sequences were checked anywhere from three to five times. The consortium plugged large DNA segments into bacterial "artificial chromosomes" (BACs), sequenced, sheared, and recloned these and finally reassembled the genome by melding all those BACs sequences together. The estimated total size of the genome is 3.2 giga (billion) bases. Only 1.1 percent to 1.4 percent of those sequences encode protein; indeed over half of our DNA is made up of boring, repeated sequences of various types: "45 percent in four classes of parasitic DNA elements, 3 percent in repeats of just a few bases, and about 5 percent in recent duplications of large segments of DNA."[14]

- There are up to about 30,000 protein-coding genes in the human genome versus 6,000 for a yeast cell, 13,000 for a fly, 18,000 for a worm, and 26,000 for a mustard seed. However, the genes are more complex, supporting the notion that we are not hard wired, with more alternative splicing generating a larger number of protein products. For example, a thirty thousand–gene organism can be made almost infinitely more complicated: with each gene interacting with four or five others on average, the human genome approaches the complexity of a modern jet airplane, which contains more than two hundred thousand unique parts.

- The genetic landscape shows marked variation in the geographic distribution of landmarks. Our "junk DNA" and the flitting, plug-in elements of DNA called "transposable elements," are not uniformly dispersed. Dozens of genes appear to have been derived entirely from transposable elements. Our master "homeobox" gene regions, which govern fetal development, show very few of these repetitive sequences. Highly expressed genes tend to be clustered in specific chromosomal regions and the so-called Alu class of repeated sequences seem to clump near the gene-rich, high GC regions, probably to turn on stress-related genes.

- Although about half of the human genome derives from transposable elements, there has been a marked decline in the overall activity of these elements in man. Man and mouse have expanded the genes that code for neuronal function, developmental regulation, apoptosis, hemostasis, inflammation, and immunity. We differ from the mouse by only three hundred genes.

- The pool of dull, repetitive sequences is a reservoir from which new functions can be drawn. Introns, those senseless sequences that break up the protein-coding exons, make up an amazing 24 percent of DNA; introns are much longer in human DNA than in the genomes of other species.

- The mutation rate is about twice as high in male as in female meiosis, showing that most mutation occurs in males. The Y chromosome is least like that of a mouse; if you want a payoff in human population genetics, the check is in the male.

- Duplications of huge chunks of chromosomes are much more frequent in humans than in yeast, flies, or worms. Centromeres and telomeres (the middles and ends of chromosomes) are filled with large, recent, segmental duplications of sequences from elsewhere in the genome.
- Hundreds of human genes probably came from bacteria at some point after our ancestors acquired backbones. Most of this parasitic DNA came about by reverse transcription from RNA.
- The full set of proteins (the "proteome") encoded by the human genome is more complex than those of invertebrates. This is due in part to the presence of vertebrate-specific protein domains and motifs (an estimated 7 percent of the total), but more to the fact that vertebrates appear to have arranged preexisting components into a richer townscape of domains. Humans do more with less.
- SNPs (single nucleotide polymorphisms) are where the real action is in pharmacogenomics, population genetics, and anthropology. Anywhere from 1.6 (consortium) and 4.0 million (Celera) SNPs have been identified. Before the genome, we had supposed that when a G became a C, or an A became a T, drastic changes would follow in our proteins. We now learn that less than 1 percent of all SNPs result in variation in protein structure.

Conclusion: This Cliff's Notes version of the genome suggests that Buffon, Lamarck, and the reductionists of the Enlightenment were correct when they drew up the family tree of man. There is, indeed, a hierarchy to biology, a great chain of being in which man is the latest link. We may be made of much the same stuff as yeast, flies, and worms, but we use that stuff in radically different ways and for different ends: no lemur has figured out its genome. The old dogma of one gene, one enzyme, one disease, may be dead. But the new, dynamic genome with its "mysteries," its "emergent" properties as Stephen Jay Gould calls them, is up for grabs by a new generation of scientists, schooled in bioinformatics, network theory, and molecular genetics.[5] We aren't talking angels. Indeed, I'd go along with Sidney Brenner who responded to a remark made by then President Bill Clinton when the two groups had their first press conference (see entry for June 26, 2000). "President Clinton described the human genome as 'the language in which God created Man.' Perhaps now we can view the Bible as the language in which Man created God."[16] My bet is that what seems like a mystery to us right now, stuck as we are in the tangled bank of the present, will yield to a new level of analysis much in the way that Darwin's notion of unspecified "laws acting around us" has yielded to the unraveling of thirty thousand human genes.[17]

LONG AGO AND FAR AWAY

All that discovery, and all that controversy. There's never been anything like it before. Or has there? Actually, there has. There is no question but that the map of the human genome is a milestone in the history of our age, not to speak of the ages. But so was the *Encyclopédie ou Dictionnaire raisonné des sciences, des arts et des métiers, par une Société de Gens de lettres* (Diderot's Encyclopedia) published between 1751 and 1772 under the direction of Denis Diderot, with seventeen volumes of text and eleven volumes of plates. Like the genome it's also on-line, thanks to the University of Chicago.[18] Contributors to the *Encyclopédie* included the most prominent philosophes: Buffon, Voltaire, d'Alembert, d'Holbach, and Turgot, to name only a few. These luminaries collaborated in the goal of assembling and disseminating in clear, accessible prose the fruits of the accumulated knowledge of their day, leaning heavily on numbers, alphabets, and sequences of practical activities. Like the genome issues of *Science* and *Nature*, the *Encyclopédie* was a book of instruction meant to be useful and practical; its organizing principle was laid out in d'Alembert's majestic first discourse: "The discovery of the compass is no less advantageous to the human race than the explanation of the properties of the compass needle is to physics."[19] There were seventy-two thousand articles—almost twice as many as we have genes in the genome—and they were written by more than 140 contributors. The entire series contains

16,500 pages on which one finds 17 million words and 2,569 plates. The whole enterprise demanded an army of manual as well as intellectual laborers; not only philosophes, but also engravers, printers, bookbinders, etc. It took over twenty years, and that was fast for the time. Diderot and company, the loosely bound Société de Gens de lettres was a private enterprise, sponsored by wealthy subscribers. This lit-tech enterprise was opposed at every turn by the king, the censors, the clergy, and most vehemently by the official Academy of Sciences.[20]

The academy had been engaged in a dawdling effort of its own for almost eighty years when it got wind of Diderot's stab at an encyclopedia. In 1675 Colbert, Louis XIV's prime minister, had asked the Academie de Sciences (which he himself had founded in 1666) "to publish a series of illustrations and explanations of the machines used in the arts and crafts." The greatest polymath of the age, René Reaumur, was put in charge; it was like asking the team of Joshua Lederberg, James Watson, and Richard Feynman to oversee the Yellow Pages. By 1759 not one fascicle had been published, although some plates had been engraved and many more drawn. When in 1759 Diderot and crew started their efforts at engraving, they looked over some of the academy's plates as possible models. But then one of Diderot's disgruntled workmen—a M. Patte—accused Diderot of taking seventy-seven plates from Reaumur's advance proofs, and the academy sent a commission headed by a mechanically astute member, the surgeon Morand, to look over Diderot's premises. They found at least forty of Reaumur's unpublished plates, and made Diderot agree to show the academy each of his new plates of the *arts et metiers* before publication to make sure that nothing was directly taken from the academy plates. Nothing ever was, and the two sides published simultaneously in the winter of 1761.

The academy's volume was called *Description des Arts et Metiers*; the first fascicle appeared eighty years after Colbert had proposed the project and ten years after Diderot et al. entered the fray. "Speed matters" proclaims Celera's motto. Speed mattered in the Enlightenment; the *Encyclopédie* became the *machine de guerre* of reductionist, secular science. The private side had goaded the public, and the verdict of history has been one that should please Craig Venter, Francis Collins, and Eric Lander: "thus one sees to what degree the emulation between the *Encyclopédie* and the Academy of Sciences, far from being harmful, was finally fruitful for both enterprises, as well as for the progress of technology in general."[21]

The scientist who would by no means have been astounded, of course, was Robert Hooke, who was exceedingly proud that the Royal Society was engaged in the sort of practical task that men of business—those with a *meum-teum* sense—could appreciate:

> But that yet farther convinces me of the Real esteem that the more serious part of men have of this Society, is, that Several Merchants, men who act in earnest (whose Object is *meum-teum* [free enterprise], that great Rudder of human affairs) have adventur'd considerable sums of Money, to put in practice what some of our Members have contrived, and have continued stedfast in their good opinions of such Indeavours, when not one of a hundred of the vulgar have believed their undertakings feasable.
>
> And it is also fit to be added, that they have one advantage peculiar to themselves, that very many of their number are men of Converse and Traffick; which is a good Omen, that their attempts will bring Science from words to action, seeing the men of Business have had so great a share in their first foundation.[22]

February 7, 2001

The Small Machines of Nature:
Life from Outer Space

A FLASH OF LIGHT IN A CLOUD OF ICE

THIS WEEK, after the mother of all earthquakes on the subcontinent killed at least twenty thousand people in Bhuj, India, a few lucky survivors offered prayers of gratitude to Mother Earth for their escape: "I don't know why this has happened to us," said Bena Ben Gopal, a farm laborer and mother of two. "I pray to the Earth Mother every day."[1] In Tripoli, celebrating the return of one of two Libyans found innocent of the Pan Am bombing, Col. Muammar el-Qaddafi slashed the throat of a camel in a ritual "sacrifice for an honored guest" and splayed the dead animal across the back of a pickup truck.[2] Meanwhile, in the realm of reason, scientists announced that they had duplicated in the lab how life on earth was assembled by a flash of light in a cloud of ice. They used ingredients in the dense, icy clouds between the stars—water, carbon monoxide, carbon dioxide, methanol, and ammonia—and irradiated them in an icy vacuum with ultraviolet light as powerful as the cosmic rays of stellar birth. And, lo, they had made a membrane, without which no life can exist.[3] The discovery, made at NASA's Ames Research Center (Moffet Field, California), is as compelling an achievement of reductionist science as the synthesis half a century ago of RNA and DNA from simple molecules by Ochoa and Kornberg.[4]

The work owes much to studies of model lipids (liposomes) initiated by
A. D. Bangham, FRS, in 1965.[5] I daresay that it owes even more to the
experimental tradition of the Royal Society, spelled out by Robert
Hooke, FRS, in 1665:

> For the members of this Society having before their eyes so many fatal
> instances of the errors and falsehoods, in which the greatest part of
> mankind has so long wandered, [they] have begun anew to correct all
> *Hypotheses* by sense, as seamen do their *dead Reckoning* by Coelestial
> Observations; and to this purpose it has been their principal indeavour
> to enlarge & strengthen the *Senses by Medicine*, and by such *outward
> Instruments* as are proper for their particular works. By this means they
> find some reason to suspect that those [phenomena] confessed to be
> occult, are performed by the small machines of Nature.[6]

This week's report in the *Proceedings of the National Academy of
Sciences* (by Jason P. Dworkin, of the Ames Center, David W. Deamer,
a former coworker of Bangham's, now at the University of California at
Santa Cruz, and Scott A. Sandford and Louis J. Allamandola, codirec-
tors of the astrochemistry laboratory at Ames) earned headlines that
were by no means exaggerated: "Chemicals Almost Come Alive; NASA
creates cell-like membranes by irradiating inert compounds."[7] "In
Space, Clues to the Seeds of Life . . ."[8] The paper itself was somewhat
more prolix. Dworkin et al. began by noting that interstellar gas and
dust are the stuff from which the solar system is formed. Once the
inner planets like earth had cooled, debris falling from outer space
probably seeded them with organic material. At the end of the early
hot phase of star and planet formation, "less refractory materials were
transported into the inner solar system in the form of comets and inter-
planetary dust particles. Delivery of such extraterrestrial compounds
may have contributed to the organic inventory necessary for the origin
of life."[9] They reasoned that interstellar ices might be the source of
compounds delivered to earth in the heavy bombardment by space rub-
ble that occurred when the earth was very young. One notes that even
today more than a hundred tons of space debris rain on Earth annually,

much of it in the form of organic material. Dworkin et al. assumed that the building blocks of comets, those masses of interstellar ice, absorbed much of the methanol, ammonia, carbon mono and dioxides in molecular clouds, dumping these on the earth sometime between the inorganic phase of the earth's history (4.5 to 5 billion years ago) and the first signs of organic life (3.5 billion years ago).

IN THE BEGINNING THERE MUST HAVE BEEN A MEMBRANE

A quarter of a century ago, based on work begun by and with Alec Bangham, it was proposed that

> In the beginning there must have been a membrane. Whatever flash of lightning there was that organized purines, pyrimidines, and amino acids into macromolecules capable of reproducing themselves, it would not have yielded cells but for the organizational trick afforded by the design of a membrane wrapping. It is in the Lucretian nature of membranes that they appear to divide cells and organelles into units of space so organized as to form the specialized reaction flasks required for biochemical processes. Were lipids, of whatever origin, to find themselves in the vicinity of primordial [synthetic] reactions, it would not be too difficult to imagine them forming self-assembled bubbles within which to segregate the new thing—life, as it were—from the hostile sea.[10]

Although the proposal was pure theory at the time, it was supported by two facts: pure phospholipid assemblies could form closed bilayer structures—just like red cells—and they could capture big molecules like enzymes and nucleic acids. Indeed, the notion was so generally accepted that it became a mantra for the National Institute of General Medical Sciences of the NIH.[11] But it was David Deamer who has been the—dare one use the term—prime mover in reducing that theory to experiment. Therefore when Dworkin told reporters that "All life as we know it on Earth uses membrane structures to separate and protect the chemistry involved in the life process from the outside," he wasn't whistling Dixie, he was echoing Deamer.[12] Deamer has rigorously pursued that notion since the spring of 1975, when he and Alec Bangham had tossed astrobiology around in a Morris MiniMinor on the A10 after Bangham had ventured on the title "Membranes Came First" for a lecture at Bristol.[13] By 1997 Deamer had formulated his own Lucretian principle that "The first systems of molecules having the properties of the living state presumably self-assembled from a mixture of organic compounds available on the prebiotic Earth."[14] Life didn't come from nothing, but from organic compounds composed of atoms, à la Lucretius:

> But in all this, when we have learned that nothing
> can come from nothing, then we shall see straight through
> to what we seek: whence each thing is created
> and what manner made, without god's help[15]

In the laboratory, Deamer had modeled the origin of life by capturing enzymes that made nucleic acids, à la Ochoa or Kornberg, in liposomes. By choosing appropriate lipids, he changed the barrier properties of the sacs so as to permit nucleic acids to accumulate in the vesicles: substrate outside, enzyme inside, RNA stayed in the sac.[16] Alas, he confessed, "Despite this progress, there is still no clear mechanism by which the free energy of light, ion gradients, or redox potential can be coupled to polymer bond formation in a protocellular structure."[17] The question, of course, is how the energy of the Big Bang (the e in $e = mc^2$) became the matter (m) of life. The liposome—which was first called the Bangasome in 1965—is part of the answer.[18]

SICUT DEI

This week's experiments held a clue as to how the free energy of light can be captured from the sun. The experimental setup at the Ames Center was designed to mimic both the vacuum and the temperature

(−441°F) of deep space. Dworkin et al. deposited gaseous mixtures (H_2O, CH_3OH, CO, NH_3) on cold aluminum or brass substrates for up to two days to form amorphous, mixed molecular ice; they then irradiated the icy mush with photons generated by a microwave-powered, hydrogen-discharge UV lamp. They next permitted the residues to warm, examined the material by light and fluorescence microscopy and studied its optical, chemical, and biophysical properties. The stuff was no longer gas, no longer ice, no longer mush: it had assembled itself into "vesicular structures," not only into vesicles, but enclosed structures that could capture other molecules in their aqueous interstices. Dworkin told reporters that they had spent months checking the experiment for error. "I was sure it was a contamination problem," he said. "But I couldn't get it not to work."[19] The vesicles fulfilled at least three characteristics of cell membranes: (a) they were composed of surface-active molecules, resembling the phospholipids in our cells and organelles; (b) they had an internal water volume that could trap dyes which could be released at will when detergents such as Triton X-100 were added; and (c) they became fluorescent—after the bubbles became laced with novel, optically active ingredients that could trap photoelectric energy, as from the sun. Each of these properties is that of a biological membrane composed of a phospholipid bilayer, or very much that of a liposome.[20]

The description of the Ames vesicles is the blueprint for a new world of living things: "And we found that our membranes grab onto other molecules that fluoresce under radiation, which might have been very useful on our own early Earth when there was no ozone layer yet to protect the earliest forms of life from solar radiation."[21] The scientists went so far as to suggest that the fluorescent molecules trapped by the vesicles could have provided a kind of sunscreen that molecules such as DNA needed for safe replication. Moreover, tested back-to-back, the surface properties of the vesicle components made in the Ames lab were similar to those of lipidlike compounds extracted from the well-studied Murchison meteorite that landed in Australia, a rock that had been the subject of an earlier Deamer study.[22] At Ames, the critical components of the residue turned out to be CO, methanol, and NH_3. No ammonia, no vesicles. But above all, carbon. The carbon had to have formed chains longer than six or eight carbons in length, the scientists argued, otherwise it would not have remained in the residues, since short-chain carbon compounds like hexane are water-soluble.[23] Chain length wouldn't have surprised Primo Levi, the best writer on chemistry of our time:

> Our atom of carbon . . . receives the decisive message from the sky, in the flashing form of a packet of solar light; in an instant, like an insect caught by a spider, it is separated from its oxygen, combined with hydrogen and . . . finally inserted in a chain, whether long or short does not matter, but is the chain of life. All this happens swiftly, in silence, at the temperature and pressure of the atmosphere, and gratis: dear colleagues, when we learn to do likewise we will be *sicut dei* [like gods].[24]

FELLOW IMMIGRANTS

These studies put a new spin on the origins of life on Earth. It had been generally accepted that life emerged from a unique, primordial soup present 4.5 billion years ago on Earth and Earth alone when it cooled. But the Ames experiments show that native precursors of organic molecules were unnecessary; immigrants arriving from space could have turned the trick. "This discovery," Allamandola told reporters, "tells us clearly that the kinds of compounds that are critically important for life's development—and that most people have always assumed are unique on Earth—can actually exist everywhere in the universe. They can fall on planets and moons, and with a little heat and a little light, chemistry can become biology."[25] The Ames experiments make it likely that the building blocks of life—those vesicles, those membranes—did not bubble up from some native, primal stew. Instead, they may have arrived much in the way most Americans arrived in the New World, as steerage passengers from teeming shores—or, in the case of vesicles, as flashes of light in a cloud of ice.

That notion would have pleased Franklin Delano Roosevelt. In

1939, shortly after the Daughters of the American Revolution had denied the stage of Constitution Hall to black soprano Marian Anderson, the president addressed the all-white, all-Protestant DAR from that same stage with the salutation, "Fellow Immigrants." It was also a flash of light in a cloud of ice.[26]

January 24, 2001

Dengue and DDT:

What Would Voltaire Do?

TO KICK OFF this week's $40 million inaugural bash, George W. Bush mounted a stage at the Texas Black Tie 'N' Boots Ball and slowly lifted a cuff of his tuxedo trousers to show black custom-made leather and suede cowboy boots embroidered with his initials and the presidential seal.[1] Meanwhile in El Salvador, a small Central American country where one in four people survives on less than a dollar a day, aid workers from a dozen countries were coping with a killer earthquake and an epidemic of mosquito-borne dengue fever. Alas, while dengue or malaria can be checked by education in mosquito control and/or the judicious use of DDT, ingrained folkways favor victory of the vector over the victim—not only in Central America. Salvadorans and Bush Republicans both called on a "pastor's prayer" for help with their special burden.[2]

WHAT WOULD VOLTAIRE DO?

On January 13 a 7.6 magnitude quake killed hundreds people in mudslides across the tiny Central American nation. Latest reports from the U.S. Agency for International Development indicated that twelve of the country's fourteen provinces had been affected by the lethal earthquake, which was followed by thirteen hundred tremors or aftershocks. At last count 707 were dead and 3,700 injured; the quake also destroyed more than 141,000 homes and forced nearly 68,000 people into shelters.[3] Hardest hit was the Las Colinas section of Santa Tecla, ten miles from the capital, where a massive mudslide flattened a new housing development, killing four hundred and leaving untold numbers of men, women, and children missing. Local authorities warned that cholera, dysentery, and dengue fever might break out in the Santa Tecla area as this large pool of homeless people, living in mosquito-ridden shelters was exposed to dirty water, stagnant puddles, and rotting flesh.[4] Twenty-four new cases of dengue fever were reported during the week by international relief teams and a Cuban aid worker had already succumbed to the disease. Relief workers rushed to locate and bury bodies pulled from the debris before fumigating disaster areas; they urged locals to bury animal and human corpses on which dengue-bearing mosquitoes feed.[5]

Fears of dengue, and especially of its lethal form, dengue hemorrhagic fever (DHF), are well founded. Indeed, all of Central America—and El Salvador in particular—has been gripped by a savage epidemic of this arthropod-borne arbovirus.[6] By December of 2000, the year's toll in El Salvador of dengue and DHF was 16,355, with 31 deaths of 336 confirmed cases of DHF. The calculated case fatality rate for DHF was 9.22 percent and the incidence rate for combined dengue and DHF was 260.60 per 100,000 population. When one considers that all of Europe is up in arms over ninety cow-borne cases of variant Creutzfeld-Jakob disease—and those in the course of a decade—one marvels at the equanimity of the Salvadorans. In fact much of Latin America has been gripped by dengue, which experts call "one of the most rapidly expanding and re-emerging infectious disease problems in Latin America. In less than 20 years, the region has transformed itself from hypoendemic to hyperendemic."[7] (Translation: a dengue-mosquito sanctuary.) In rural areas of Central and Latin America, where dengue takes its greatest toll, natives do not believe that mosquitoes carry disease. Bitten day and night by one or another insect, disbelief in the infectious nature of dengue permits them to work in the field and sleep in unprotected huts. They remain convinced that dengue fever is a sign of divine displeasure: "I've seen it many times in the people in my church," said the Rev. Ernesto Perez, an evangelical pastor. "The disease signifies that Jesus Christ is testing people. It is a test not everyone can survive."[8]

I'm not sure that we in the developed countries can remain smug about this widely held popular belief; citizens of the United States have just recovered from an electoral campaign marked by a pitch of piety not sounded in our country since the days of William Jennings Bryan. When asked to name his favorite political philosopher, the Republican presidential candidate had replied: "Christ, because he changed my heart." What came to Al Gore's mind when faced with a difficult question? The initials W. W. J. D., that's short for "What would Jesus do?"[9] Since the major act of God this week was a terrible earthquake that killed hundreds of men, women, and children, one might ask W. W. V. D.? Voltaire was, of course, prompted to write *Candide* (1759) by the great Lisbon earthquake of November 1, 1755, a disaster that killed thousands of innocent children and that literally shook the philosopher's faith in divine providence forever:

> After the earthquake which had destroyed three-fourths of the city of Lisbon, the sages of that country could think of no means more effectual to preserve the kingdom from utter ruin than to [burn] a few people alive by a slow fire, and with great ceremony, as an infallible preventive of earthquakes. . . . Candide, amazed, terrified, confounded, astonished, all bloody, and trembling from head to foot, said to himself, "If this is the best of all possible worlds, then what are the others? (Si c'est ici le meilleur des mondes possible. Que sont donc les autres?"[10]

FLYING SYRINGES

The epidemiology and pathophysiology of the dengue viruses have been well studied. Originally endemic in Southeast Asia, they are now found almost everywhere malaria is rampant. A variety of mosquitoes carry the four serotypes of dengue viruses (D_1-D_4), which differ not one whit in their capacity to cause dengue fever, dengue hemorrhagic fever, or dengue shock syndrome. Central America has not only become a way station in the drug traffic to North America but also a reservoir for proliferating dengue viruses.[11] Over the last twenty years, serotype circulation in the region has gone from none or single to multiple, and the mosquitoes are moving north. Our porous southern border has permitted dengue to become endemic in southern Texas and the disease is now moving into California.[12] "Mosquitoes are flying syringes," warned Dr. Cortez-Flores of Loma Linda School of Public Health, an expert on the arthropods.[13] As Steven Soderbergh's epic film *Traffic* demonstrates, syringes know no borders. The notion, I might add, will be familiar to students of the slave trade. *Aedes aegypti*, a prime vector of dengue and yellow fever, reached South America in slave ships on which the mosquitoes' eggs survived in water containers while live mosquitoes fed on the populations on board. Michael Nathan, a WHO entomologist, believes that "A lot of slaves and a lot of crews died of yellow fever on the way over."[14]

Severe dengue fever is an immune complex disease: we are made sick not only by the virus, as by the flu, but by our own antibodies to the virus.[15] A first infection by dengue virus is usually mild, almost flulike; but the severe forms, which include "breakbone fever" with its triad of fever, rash, and acute muscle pain, usually follows the second exposure to the same or any other serotype. The crueler syndromes—DHF and hypovolemic shock—tend to affect younger patients, with mortality rates up to 60 percent.[16] The disaster of dengue hemorrhagic fever is caused by immune complexes with the virus as antigen tagged by our own antibodies. These in turn provoke clumping of platelets and white cells in the circulation; when these aggregates get stuck in the blood vessels of lung or kidney, organ damage results. Worse yet: the blood vessels are injured by chemical messengers of inflammation (complement-derived anaphylatoxins, C3a, C5a). Acute respiratory distress syndrome and disseminated intravascular coagulation follow, as activation of endothelial cells and macrophages floods the system with procoagulant factors and inflammatory cytokines (TNF, RANTES, IL-8); fibrinolysis, vascular leakage, hemoconcentration and—alas—exsanguination are terminal events.[17] There is no specific treatment other than hydration and/or transfusion; steroids are useless.

WHAT WOULD RACHEL DO?

We cannot treat dengue, but we now have a new weapon to prevent most mosquito-borne diseases: DDT. Yes, DDT is back. Last month, at a meeting in Johannesburg, officials of the United Nations Environment Program agreed to permit use of DDT in malaria or dengue control programs.[18] They were responding to almost a decade of worldwide clamor by physicians, public health officials, and malaria experts at the World Health Organization that DDT was a necessary public health weapon in poor tropical countries. The arguments have been summarized by Amir Attaran of Harvard University's Center for International Development.[19] Noting that four hundred physicians the world over had signed a petition urging resumption of DDT spraying for mosquito elimination, Attaran pleaded that DDT house spraying "is an inexpensive, highly effective practice against malaria...the quantities involved are minimal ($2g/m^2$) unlike agricultural uses which inject tons of DDT into the outdoors." Most tellingly, when extrapolated to local disaster areas like Santa Tecla:"For the amount of DDT used on a cotton field, all the high-risk residents of a small country can be protected from malaria."

Medical opinion has come full circle on the DDT issue. DDT was first introduced to the world by Swiss chemist Paul Hermann Müller (1899–1965) of Geigy AG who passed its secret to the Allies towards the end of World War II. Wherever this pesticide was used, insect-borne diseases were essentially eradicated, especially the flea-borne scourge of the camps, typhus. The liberators of Buchenwald carried DDT in their jeeps. Müller won the Nobel Prize in physiology or medicine in 1948 for his "discovery of the high efficiency of DDT as a contact poison against several arthropods." The full citation gives him proper credit: "DDT has been used in large quantities in the evacuation of concentration camps, of prisoners and deportees. Without any doubt, the material has already preserved the life and health of hundreds of thousands."[20] Shortly thereafter, DDT became available worldwide and was successfully used to eradicate malaria from the developed regions of the world (the United States, Europe) and to lower case rates by over 99 percent in others (Sri Lanka, India). In South Africa's KwaZulu-Natal Province, for example, malaria epidemics before DDT killed more than twenty-two thousand. After health authorities began spraying DDT, the incidence dropped dramatically. By 1973, South Africa recorded only 331 malaria cases in the entire country and in 1977 only a single death.[21]

And then along came Rachel Carson (1907–1964). A fine naturalist, and longtime marine biologist at Woods Hole, she called general attention to the effects of DDT on the eggs of raptor birds and seashore life. She won public acclaim for her splendid prose and public sympathy for her poignant situation. Carson's plea for an end to chemical pollutants was written as she was dying of breast cancer. Chemicals, she wrote "have the power to kill every insect, the good and the bad, to still the song of birds and the leaping of fish in the streams, to coat the leaves with a deadly film, and to linger on in the soil. . . . Can anyone believe it is possible to lay down such a barrage of poisons on the surface of the earth without making it unfit for all life? They should not be called 'insecticides' but 'biocides.' "[22] By the mid-seventies WHO dropped its DDT program, having been convinced by Carson's followers that her poetic hunches were supported by robust data in the field. Indiscriminate use of DDT had led to pollution of soil and stream and ocean. By then, millions of tons of DDT had been spread over farmland and forest, suburb and seashore.

But the ban on DDT had unintended consequences in poor countries. Thanks to DDT, countries such as Zanzibar had reduced the percentage of their populations infected with malaria from 70 percent in 1958 to under 5 percent in 1964. When the DDT spraying was halted, the malaria rate rose back to over 50 percent by 1984.[23] Swaziland, which did not halt DDT spraying, maintained its malaria infection rates between 2 and 4 percent, while just forty miles away, South Africa, which banned DDT in the '80s, watched malaria infection rates rise to 40 percent. This year South Africa resumed spraying, and after the Johannesburg meeting the UN went along. DDT is back.[24]

A final note. Carson and many of her followers believed that DDT was in some fashion responsible for breast cancer—and early work by Mary Wolff of Mt. Sinai had in fact suggested such.[25] Not so: eight studies, including reliable surveys from Sweden, Denmark, the U.K. and the U.S., have shown no direct relationship between DDT exposure and breast or endometrial cancer, although similar studies have shown the risks of other "endocrine-disrupting pollutants" such as PCBs.[26,27] Harvard's Attaran argues that "Very few other chemicals have been given such extensive scrutiny, and there is still no epidemiological or human toxicological evidence to impugn DDT."[28] And the latest, just in, is also by Mary Wolff, who—like any good scientist—seems to have changed her mind, convinced by her own evidence that levels of DDT, or its metabolite, DDE, are *not* associated with risk for breast cancer.[29] Rachel Carson, no mean scientist herself, spent the last years of her life shuttled between doctors who tried to stem her metastases with everything from radiophosphorous and Krebiozen.[30] She would have been proud of Ms. Wolff's devotion to the facts of breast cancer epidemiology—and of WHO efforts to prevent children from dying in mosquito-ridden camps.

January 17, 2001

Unspoken Issues

STUDENTS, FACULTY, house officers, and attending physicians at the NYU School of Medicine are grouped into clinical firms and collegial societies named after distinguished NYU physicians such as H. Sherwood Lawrence, Saul Farber, Lewis Thomas, et cetera. This week a fledgling member of the Lewis Thomas Society called me up to prep him on the life of a man he'd never met. He knew that I had worked for many years with Lewis Thomas and asked me to flesh out a short biographical sketch he'd written to inform the new generation of who Thomas was and what he had done. Memory is not always the best guide to fact, so I turned to Lew's autobiography and came up with a brief outline of that very fine life.

The life of Lewis Thomas (1913–93) spanned the golden age of American medicine, an era when—as Thomas put it—our oldest art became the youngest science. Thomas played a major role in that transformation and became known among scientists as an innovative immunologist and medical educator. He became far better known as a deft writer whose essays bridged the two cultures by turning the news of natural science into serious literature. Witty, urbane, and skeptical, he may have been the only member of the National Academy of Sciences to have won both a National Book Award and an Albert Lasker Award.

He is certainly the only medical school dean whose name survives on professorships at Harvard and Cornell, a prize at Rockefeller University, a laboratory at Princeton, and on a book that is eleventh on the Modern Library's list of the best one hundred nonfiction books of the century.

On another level, however, the lifetime of Lewis Thomas coincided with a special period in American medicine, a time when its scientific base became the strongest it had ever been and its social impact the greatest. Indeed, judging from the numbers who came from overseas to learn from it, American medicine became the envy of the world. But as the balance shifted in medical science from the old world to the new, doctors went out of the business of laying on hands and took on the job of clicking out genes. It was not by accident but by design that American medicine was turned from a nineteenth century folk art into a twentieth century science. After the Flexner report of 1910, medical instruction became largely concentrated in university hospitals where the modern sciences of immunology, biochemistry, and genetics could be pursued as eagerly at the bedside as in the lab. Lewis Thomas and his generation of immunologists presided over the conquest of polio and rheumatic fever, the achievements of blood banking, cardiac surgery, and the transplantation of organs, not to speak of the discovery that DNA was the basic unit of genetic information.

Thomas and his colleagues were educated in colleges at which the liberal arts were still firmly in place and after John Dewey's learning-by-doing had moved from the schools into the universities. It was an era when those who did medical science were expected to know why it was done and for whom. There was, however, one issue they dared not broach. That was the issue of who they were, and who was permitted to join their ranks. I stumbled across this issue spelled out in a striking passage describing Thomas's arrival at NYU after an education at Princeton, Harvard Medical School, and faculty positions at Columbia, Tulane, and the University of Minnesota:

[NYU] was also known to be a school largely and traditionally populated by students from New York City itself, many of them from relatively poor families, mostly Jewish, some first-generation Italians, a few Irish Catholics, a very few blacks—a different student body from those at Columbia or Cornell.[1]

What Thomas meant, of course, is that unlike NYU, the medical schools of Columbia and Cornell had a strong Jewish quota in place since the 1920s. And while Thomas was probably as indifferent as anyone to questions of race or religion when it came to his professional associations, the schools he attended and the institutions in which he worked were steeped in resolute traditions of de facto and de jure discrimination. While no one would be astonished that from the 1920s to the 1940s Princeton totally excluded blacks, and while Yale and Harvard admitted only a handful, those barriers extended past the Ivy League.[2] For example, it was not until 1959 that Lewis Thomas, as one of his initial appointments, named the first identifiable Jewish chief medical resident of the III (New York University) Medical Division of Bellevue Hospital.[3] Thomas's predecessor, William S. Tillet, although recognized as a distinguished educator and researcher, had been no obvious philosemite: in a school whose student body was between 60 and 70 percent Jewish, the percentage of Jews on the house staff of Tillett's IIIrd Division never exceeded 20 percent. All that changed when Lewis Thomas arrived; it was a small signal that the guard was changing.

Before the Civil War—indeed, until the great migration from Eastern Europe at the end of the nineteenth century—the few well-off Jews in the United States, who hailed mainly from Germany, were by and large well tolerated at elite universities.[4] That unstable equilibrium was upset by the influx of a new class of poor, hardworking immigrants from Poland and Russia whose children pressed American universities for admission. Their academic skills posed a problem in squaring the values of class and conduct with those of education; the threat was that the "Jewish problem" might become the the "Jewish invasion."[5] By the end of World War I, the invasion had spread from public to Ivy League campuses: in 1918 80 percent of the students at New York's City College and Hunter College were Jewish, as were 40 percent of the students at Columbia. A jingle circulated on New England campuses

> Oh Harvard's run by millionaires
> And Yale is run by booze,
> Cornell is run by farmer's sons
> Columbia's run by Jews.[6]

But the campuses of millionaires and boozers were also subject to the Jewish invasion. Between 1900 and 1920 the percentage of Jews at Yale increased from 2 to 10 and the freshmen enrollment of Jews at Harvard had risen from 7 to 22 percent. President A. Lawrence Lowell complained of the "Little Jerusalem" that formed itself in Walter Hastings Hall where a large number of Jewish undergraduates lived, and Samuel Eliot Morison explained why Harvard found this unsuitable: "The first German Jews who came were easily absorbed into the social pattern; but at the turn of the century the bright Russian Jewish lads from the Boston Public Schools began to arrive . . . and in another fifteen years Harvard had her 'Jewish problem.' "[7]

The Association of New England Deans met to solve the problem. Kenneth C. M. Sills of Bowdoin openly admitted that "We do not like to have boys of Jewish parentage," while Frederick S. Jones of Yale complained that they were "likely to overrun us. A few years ago every single fellowship of value was won by a Jew. . . . We must put a ban on Jews." And in 1920, Dean Otis E. Randall of Brown argued that this could only be effected by "limitation in the enrollment of Jews and Negroes." The deans of every elite institution in the Northeast proceeded to limit enrollment by simply imposing more or less explicit quotas on Jews, blacks, and Asians. (Women, of course, were tucked away in the "Seven Sisters.")

The quota system was in keeping with the nativist policies of the federal government, which in response to postwar isolationism had passed the strict immigration laws of the '20s that excluded folks from eastern Europe, the Mediterranean, Asia, and Africa. Lowell, who had become vice president of the national Immigration Restriction League six years after becoming president of Harvard, wrote that the federal government had faced "the problem of mass immigration by imposing a system of quotas"; a similar quota system would ensure "tolerable homogeneity" at its most prestigious university.[8] So it did: Lowell and the league made sure that by the mid-twenties, entry to campus as well as to the country was determined by national or religious quotas. Quotas, more or less rigorously enforced for much of the twentieth century, guaranteed that Lewis Thomas would find at the NYU School of Medicine "a different student body from those at Columbia or Cornell."

Lowell had argued in 1922 that quotas would limit the number of "let us say Orientals, colored men and perhaps French Canadians" who did not "mingle well" and that limiting the percentage of Jews at Harvard to 15 percent would stem the "growth of antisemitic feeling" in the Harvard community. In a like vein of benevolence, Lowell justified the exclusion of blacks from the freshmen dining halls of Harvard by contending that integration would "cause a revulsion and reprisals in a good many places against Negroes."[9] Lowell and his fellow university presidents did their part for bonhomie and mingling well by changing their admission practices to favor "geographic distribution," "extracurricular interests," and "mental alertness" over mere academic achievement. Harvard's application for admission now asked, "What change, if any, has been made since birth in your own name or that of your father. (Explain fully.)" The result of these practices was that in the years between 1922 and 1926 Columbia's percentage of Jews was reduced from 40 to 20 percent Yale's from 14 to 10 percent, and Princeton's already meager portion from 4 to 3 percent.

During the years of Thomas's undergraduate education, discrimination was guaranteed by an only slightly disguised quota system, which was often simply enforced by not granting interviews to graduates of the public high schools in the immediate surroundings of Princeton, Harvard, or Yale. Hearing that most of the Jewish undergraduates at Yale had come from New Haven, Bridgeport, or Hartford, President Angell suggested in 1936 "That, if we could have an Armenian massacre confined to the New Haven district, with occasional incursions into Bridgeport and Hartford, we might protect our Nordic stock almost completely."[10]

The Nordic stock preserved itself at Harvard, but only after four years of bitter controversy and appeals. Strong opposition arose to the first overt imposition of quotas by Lowell: liberals rallied around the emeritus president of Harvard, Charles W. Eliot. Among their number were the deans of the law and medical schools, Roscoe Pound and David Edsall, alumni such as Walter Lippmann and Judge Learned Hand. Once the issue became public, they were joined by politicians such as Mayor Curley of Boston and labor leaders such as Samuel Gompers. Overt quotas were rejected by a faculty more inclined to tolerance than its president, and Lowell felt obliged to resort to indirect measures. As part of the House plan he instituted at the College, admissions were limited to a thousand each year and it was agreed that the top 7 percent of graduates from a "selected" number of secondary schools would be admitted automatically. It was therefore simple to juggle the ethnic mix of admissions by excluding from the schools selected those from which the bulk of Jewish applicants came. Yale followed suit and, in the words of its dean of admissions, "drew a doughnut around the suburbs of New York, Bridgeport and New Haven, the high schools of which were less than entirely Nordic; Jewish faculty members complained that the ring was a bagel."

In Cambridge, Lowell's various reforms made it possible to meet covertly the overt aim of Harvard's dean, Henry Pennypacker, to reduce the "Hebrew total to 15% or less by simply rejecting [them] without detailed explanation." Since the dean could not effect this reform de jure, he accomplished it de facto by imposing regional quotas, and by making "character" or other secondary social qualities important criteria for undergraduate admissions. Key to it all was the personal interview, carried out by Pennypacker and his colleagues, against which Harry Wolfson—then an assistant professor of Jewish literature— argued "It may be readily admitted that outward appearance is a proper test for selecting book agents, bond salesmen, social secretaries and guests for a week-end party, but scarcely a proper test for the selection of future scholars, thinker scientists and men of letters."

In the 1960s those sentiments were echoed by Lewis Thomas's finest appointment to the NYU faculty, Nobelist Baruj Benacerraf—who in the days of the quota system had been rejected by every American medical school except the lowly regarded Medical College of Virginia. Addressing a meeting in Williamsburg of those selected for the highly competitive awards of the Arthritis Foundation, he told the selection committee: "Don't have an interview. It only rewards table manners. Judge only what has been written or discovered!"[11]

Lowell's innovations reduced the number of Jews admitted to Harvard from a high of 25 percent in 1925 to 16 percent in 1930, and pari passu the percentage of Jews at the Medical School declined from 16 to less than 10. There was no comparable number of blacks to speak of. In their history of medicine at Harvard, Henry K. Beecher and Mark D. Altschule reported that "since our arrival at Harvard Medical School in 1928, [we can] attest to the desire of the School to include blacks in its student body. One or two were generally on hand."[12] Women were excluded from the Harvard Medical School until World War II, when there were fewer suitable male applicants.

Nevertheless, Dean Edsall was a strong voice in opposition to discrimination in any form at the medical school. Allied to the more liberal Eliot faction, he had already battled in 1919 with conservatives over the appointment of Alice Hamilton to the medical school faculty as an assistant professor of industrial medicine.[13] Himself a pioneer in public and industrial health, Edsall assured her that she was preeminent in her field and that she deserved the job because

of the service that you have rendered to the country already, although I think it is the first time that the proposition has ever come up to have a woman appointed to any position professorial or other in the University. Aside from my very great desire that [you accept this position] I desire it also for the reason that I think it would be a large step forward in the proper attitude toward women in this University and in some other Universities.[14]

Alas, he was never able to prevail with his faculty—nor with his university presidents, Lowell and James B. Conant, to admit female students to the Harvard Medical School. Nor could he persuade them to permit Alice Hamilton to join the Harvard Club, to walk in the academic procession—or to obtain faculty seating at the Harvard-Yale game. Antifeminism was not limited to those petty matters: In contrast to the University of Pennsylvania, which admitted women to its medical school in 1914, or to Columbia and Yale, which took the "large step" in 1917, the Harvard Medical School first admitted women in 1945. Joseph Aub, a protégé of Edsall's, a colleague of Alice Hamilton's in public health, and a pioneer of the study of lead poisoning, recalls the reluctant acceptance by the medical school faculty of his wartime proposal to admit women. His motion carried, but only after there had ensued a "serious commotion; one of the [opposing] professors at the meeting developed a paraplegia that night."[15]

In the tradition of Aub and Edsall, Lewis Thomas took the liberal side in many such discussions. In the spring of 1980, at a meeting in New Haven of the Interurban Clinical Club—that invisible college of professors of medicine from Boston, New Haven, New York, Philadelphia, and Baltimore—I proposed for membership the distinguished medical geneticist, Rochelle Hirschhorn, also of NYU. I had discussed the plan with Thomas in advance; he advised me that it would be "tough sledding." He reminded me that William Osler— by nature wary of women in medicine—had founded the club (1905), that some Hopkins traditionalists, who showed up at meetings sporting club ties that advertised Osler's magnum opus *Aequanimitas*, would find some way of scuttling the nomination. "But, what the heck, let them grumble," said Thomas, "simply tell them you can't do genetics without that extra X [chromosome] around." In the event, the vote was taken, and although a few members from Baltimore abstained (one murmured that Osler would turn over in his grave), a woman was elected unanimously. No reports of paraplegia followed; in 1987 Rochelle Hirschhorn became president of the Interurban Clinical Club.[16]

Throughout Thomas's years at the Harvard Medical School, the percentage of Jewish medical students never exceeded 10 percent.[17] Nevertheless, the liberal Edsall had appointed a good number to the faculty, including his colleagues in industrial health, Joseph Aub and Milton Rosenau, as well as Hermann Blumgart, Samuel A. Levine, Soma Weiss, and Maxwell Finland in medicine, Edwin Cohn in biochemistry, and Leo Alexander in neuropathology. The better angels of Harvard's nature won out over Lowell's nativism when Soma Weiss (born in Hungary) was named Hersey Professor of the Theory and Practice of Medicine and chief of the Medical Service of the Peter Bent Brigham Hospital in 1939, when Maxwell Finland (born in Russia) became George Richards Minot Professor of Medicine and director of the Second and Fourth (Harvard) Medical Services and director of the Thorndike Memorial Laboratory at Boston City Hospital in 1963, and when Weiss's successor, George W. Thorn, became the Samuel A. Levine (born in Poland) professor of medicine at the Brigham. Most of these appointees of Edsall figure prominently in histories of medical science, while Blumgart, Weiss, and Finland were especially important to Lewis Thomas. Leo Alexander, on the other hand, is nowadays chiefly remembered as the subject of a student poem by Lewis Thomas. Alexander was an experimental pathologist who kept a colony of pigeons permanently addicted to alcohol, hoping to show that sentient birds—like the chronic alcoholics of Boston—would develop discrete neurological and behavioral deficits. Thomas penned a poem to Alexander at a celebratory dinner for the pathologist:

> Hail to Alexander, that giant Man of Science,
> And all the happy pigeons, his little waddling clients,
> A happy lot, I truly wot, to meditate upon—
> Each little bird, with vision blurred may daily tie one on.
>
> May daily have his highball, with little ice cubes clinking,
> May walk around his cage all day while absolutely stinking,
> May peer upon the moving world with slight nystagmus trouble,
> May know the rich experience of seeing Leo double![18]

And when Thomas attained positions in which he was able to pick his own associates—black or white, male or female, Gentile or Jew—he followed the egalitarian precedent of Edsall: the range of his appointments guaranteed that everyone would have the rich experience of having the likes of Leo—or other "minorities"—as colleagues. The schools have changed and the numbers are impressive as well. By 2000, white males (including Irish, Italians, Jews, etc.) constitute only 23 percent of the Harvard Medical School students, Asian Americans (both males and females) 25 percent, and African Americans (male and female) 11 percent. Like most of the world, Harvard has changed.

January 10, 2001

The Balkan Syndrome

THIS WEEK Europe turned from its preoccupation with mad cow disease, a disease for which the Brits took the rap, to a new syndrome for which they blamed the United States. On January 4, Romano Prodi, president of the European Union's executive commission, requested that the EU investigate whether depleted uranium (DU) from U.S. ammunition had caused death and illnesses among peacekeeping forces in Bosnia and Kosovo.[1] Prodi acted in concert with Italian prime minister Giuliano Amato, who assured the tabloid *La Repubblica* that concerns about the new "Balkan syndrome" were "more than legitimate. We've always known that [depleted uranium] was a danger only in absolutely exceptional circumstances such as picking up a fragment with a hand with an open wound, which in normal circumstances isn't dangerous at all. But now we're starting to have a justified fear that things aren't that simple."[2] Nothing is simple when uranium hits the headlines.

American forces were reported to be the only NATO troops to use DU ammunition in the Balkans: more than thirty-one thousand rounds were fired. DU ammo is made of uranium 238, which is the densest metal in our arsenal and easily penetrates armor without mushrooming on impact. U 238 (half-life > 4.5 billion years) is the residue of raw uranium ore from which gamma ray–emitting U 235 has been extracted; DU has only 40 percent of the native ore's radioactivity, emitting chiefly alpha rays which do not carry past their point of contact. Nevertheless, panic gripped the EU this week when it was reported that a UN site-inspection team had found low levels of beta radiation in eight of eleven sites in Kosovo.[3] According to Pekka Havisto of Finland, who headed the UN team, an undetermined number of DU ammunition fragments remain underground at these sites; he worried that residues might leach into the groundwater. Worse yet, if the sites are cleared by detonation, aerosolized microspheres could be released. But that's for the future. Right now, Italy is particularly concerned about what happened in 1999. More than fourteen thousand rounds of ammunition launched by American A10 bombers fell in the area of Kosovo now controlled by Italian troops, according to the Italian deputy ecology minister, Valerio Calzolaio.[4] Thirty veterans "contracted serious illnesses"; twelve of these developed "cancer" while six Italian soldiers have died of leukemia (AML) since returning from the Balkans.[5] If true, the incidence exceeds that expected worldwide (one per a hundred thousand at ages fifteen to thirty-five) but a proper analysis would require a control population of the Italian military *not* deployed in the Balkans.[6] Perhaps the most convincing case against the DU-leukemia hypothesis was made by hematologist Eric Wright of Dundee: "The diagnosis of leukemia in many of these people is [too] soon after the alleged exposure." The well-documented period between exposure to radiation and development of leukemia is two to ten years.[7]

POP GOES THE SYNDROME

The latest Italian casualty was Salvatore Carbonaro, twenty-four, from Sicily, who died in November after serving twice in Bosnia but never in Kosovo, where the bulk of depleted uranium had been used. The Italians claimed they had never been told that DU was used in Bosnia until NATO 'fessed up this week "We've always known that [depleted uranium] was used in Kosovo, but not in Bosnia."[8] Leukemia was also reported in four French, two Dutch, and two Spanish peacekeepers, but the Balkan syndrome eludes ordinary medical nosology since its effects are not limited to the blood.[9] The syndrome was given its name in Belgium after nine of the country's troops who had been in Bosnia and other parts of the former Yugoslavia fell ill with various forms of "cancer." Five subsequently died, according to Belgian press reports. "Other soldiers who had been on Balkan peacekeeping missions during the 1990s reported a variety of unexplained ailments, including headaches and insomnia," reported a Belgian spokesman in December.[10] Other features were added when *La Repubblica* reported the first case of the Balkan syndrome in the U.K. Featured on its front page was a bald Kevin Rudland, a former Royal Army engineer (reserve) with full-body tattoos. Rudland had served in Bosnia for six months in 1996 and was now suffering from "stanchezza cronica e osteoartrite, perdita dei denti e dei capelli, gravi problemi intestinali" (chronic fatigue, osteoarthritis, loss of hair and teeth, and grave intestinal problems). Although his doctors are treating him with "psicoterapia" (psychotherapy), the Italians believe that he is "il primo soldato inglese colpito dall'uranio impoverito usati nei bombardamenti." (Balkan syndrome).[11] Rudland was more expansive at home: he told the *Guardian* that in addition to chronic fatigue syndrome he was suffering from "post-traumatic stress disorder."

In Belgium, the pivotal case was that of Commandant Frank Cop whose "career was cut short by illness two months after returning from duty as a peace monitor with the UN and the EU during the Bosnian war." For five years Cop claims to have suffered from "headaches and muscle aches, debilitating lethargy, and skin complaints so serious he finds it uncomfortable to bathe." The *Observer* of London chimed in that Cop and Rudland are victims of a "mysterious 'Balkans Syndrome', similar in its symptoms to the Gulf War Syndrome claimed by veterans of the war against Iraq ... most prominent among them chronic fatigue."[12] Not to be outdone by mere men in the symptom quest, French journalist Marie-Claude Dubin testified before a government probe into DU-induced illness that she had suffered a wide range of complaints since covering the Persian Gulf and Balkans. "I'm not saying that uranium was certainly the cause, but I am ill and I have the same symptoms as the veterans," Ms. Dubin told the daily *Le Parisien*. She was more dramatic in a Reuters interview given at her Paris apartment: "They sent us to the slaughterhouse." She wasn't exactly reporting from Dachau.

From mid-December on, official investigations had been launched by France, Spain, Portugal, Belgium, the Netherlands, and Finland to determine whether DU weapons had anything to do with the Balkan syndrome. Only Belgium and Spain offered early statistical analysis: defense officials maintained that the nine cancer cases, of twelve thousand Belgian troops in the Balkans and the two Spaniards, of thirty-two thousand, fell within the range expected among age-matched civilians.[13] Nevertheless, fueled by headlines that charged U.S.-led NATO forces with using uranium to provoke leukemia—not to speak of chronic fatigue, headaches, muscle aches, and "grave intestinal ailments"—the Balkan syndrome was on its way to joining the Gulf War syndrome on the list of twentieth century plagues spread by Uncle Sam.

POLITICALLY DEPLETED URANIUM

DU has long been suspected of being involved in the elusive Gulf War syndrome. Based on this premise, an English biologist and rabid anti-DU spokesman, Dr. Roger Coghill, predicted at the beginning of the air war in 1999: "We think that in the Kosovo conflict, as a result of [the use of depleted uranium by NATO forces], that there will be some 10,000 deaths from cancer."[14] His scary warnings include the notion that "one single particle of depleted uranium lodged in a lymph node can devastate the entire immune system." The first of these sentiments has been dismissed by arms expert William Arkin as "asinine" because it would postulate one death from cancer for every three DU rounds fired: since the U.S. fired 850,000 DU rounds in the Gulf War, over a quarter of a million cases of cancer should have been diagnosed.[15,16] Arkin points out that "Further, there's no proof that a single American soldier is sick purely as a result of DU exposure." His assertion is supported by studies from the U.S. Department of Veterans Affairs which has been monitoring twenty-nine American victims of "friendly fire" in the Gulf who still have DU fragments embedded in their bodies and who show higher than normal levels of uranium in the urine.[17] Compared to thirty-eight age-matched controls, their reproductive health has been normal and babies have shown no birth defects. A statistical relationship *was* observed between urine uranium levels and lowered performance on computerized tests assessing performance efficiency, but this was not a blinded study. Coghill's immunologic fantasy can be put to rest as well; every radiotherapist who has treated lymphomas knows that gated doses of 40–45 Gy of gamma radiation (one rad = 0.01 Gray) are needed to "devastate the entire immune system," that beta emission simply cannot do the job, and that airborne alpha radiation cannot cause leukemia.[18] Arkin admits that facts will not persuade the anti-DU community, and argues that perhaps the defense establishment should not use DU anymore, because it is a "politically depleted weapon." As airborne ammunition, it's old hat these days, and better ground ammunition should be worked out as well. "But that's the DU debate. The anti-crowd radiates, the press propagates, and the military goes into defilade."[19]

DEPLETED URANIUM IN THE LAB

Is there any evidence that DU *can* produce disease at all? Answers are equivocal. In 1999 scientists at the U.S. Armed Forces Radiobiology Research Institute in Bethesda implanted shell fragments of DU or tantalum (as a control metal) directly into the brains of rats.[20] Six months later, synaptic potentials in DU-exposed tissue were diminished while at twelve months they were significantly increased. These changes were no longer evident at eighteen months. Conclusion: young rats with DU implanted in their skulls have impairments that vanish with age, but they found no evidence of cancer, leukemia, or immune deficit. What about tumors? In whole animals, DU—as opposed to raw uranium—does not produce tumors in the respiratory passages (so much for Mr. Havisto). When enriched or depleted uranium was instilled into the respiratory tracts of rats, "squamous cell carcinomas developed at the site of deposition of the U 235 enriched uranium oxide in many cases but no lung tumors occurred in the rats with the depleted uranium U 235, in which the lung tissue was exposed to very few fission fragments."[21] Studies in vitro showed that 1 mg/ml of DU can indeed provoke neoplastic changes in cultured cells, but that the risk of cancer induction from internalized DU exposure was comparable to "other biologically reactive and carcinogenic heavy-metal compounds (e.g., nickel)."[22] It is of course highly unlikely that humans could be exposed to a milligram of DU from aerosolized microdrops.

THE GULF OF COMPENSATION

Neither Europe's worries nor ours have been quieted by soothing words from the Pentagon. Lt. Col. Paul Phillips told the world press that the United States "has conducted many studies on depleted uranium, particularly since the 1991 Persian Gulf War when the weapons were first used." In each case no evidence was found that depleted uranium presented significant or residual environmental or health risks, a judgment echoed by NATO spokeswoman Simone de Manso in Brussels.[23] Europe remained unconvinced; its press went bonkers. Amused that we had elected a president who had never visited their continent, America was in the dock not only for ignorance but arrogance. Labor in Britain, greens and communists in Italy, socialists in Spain, and centrists in Portugal—all demanded inquiries, commissions, apologies, and reparations.[24] The critics pointed to Gulf War syndrome as an example of DU-induced disease; nothing could dissuade them. Arkin had it just right: "In the DU world, for every crackpot haunted by radiation, there is a craven and unsympathetic commander or bureaucrat. In the middle are many physicians and suffering veterans who think DU is the cause of the post–Gulf War nightmares many veterans suffer. They are not soothed by reassuring studies and presidential commissions."[25]

Well, let me try to help. I'm not sure that there is such a thing as the Balkan syndrome, and, despite our own pusillanimous response to the Gulf War, I'm dead sure that there is no such thing as the Gulf War syndrome. Reliable, conclusive epidemiologic studies have shown that there is no difference between the symptoms reported by nondeployed U.S. servicemen and women and those who served in the Gulf. The same complaints were present in both groups: backache, insomnia, skin rashes, chronic fatigue, GI distress. "Identification of the same patterns of symptoms among the deployed veterans and nondeployed controls suggests that the health complaints of Gulf War veterans are similar to those of the general military population and are not consistent with the existence of a unique Gulf War syndrome."[26] Deployed veterans simply complained louder and in greater numbers.[27] In return, the Gulf War syndrome vets were given compensation payments by their government, a situation familiar to Commander Cop of Belgium and Kevin Rudland of Britain. The syndrome may be real in the mind of patient, physician, politician, or journalist; but I doubt that it is real in the way that AIDS, Ebola, or BSE are real.

I'm afraid that I agree with the medical historian Edward Shorter who found a unifying thread to all those "support groups" for fibromyalgia, chronic fatigue syndrome, and chronic Lyme disease. He points out in "From Paralysis to Fatigue" that "Although the amplification of normal bodily symptoms and phobias about disease have existed in all times and places, it is this delusional clinging to the belief in a given illness, that marks the last decades of the twentieth century." Shorter has convinced me that social templates shape medical fashion and that medical fashion shapes the symptoms that patients select.[28] Those symptoms—such as fatigue, intestinal complaints, insomnia, et cetera—could, of course, be produced by organic disease; that's exactly why they tug so hard at our diagnostic sleeve. The victim of the Gulf War or Balkan syndrome is in very real pain—but of the mind, and the mind chooses symptoms that will be taken as evidence of real physical disease and that will win the patient an appropriate response and compensation; "Thus most of the symptoms... have always been known to Western society, although they have occurred at different times with different frequencies: Society does not invent symptoms; it retrieves them from the symptom pool." My best guess is that the Balkan syndrome, like the Gulf War syndrome, will join the roll of stress-related, compensation-rewarded, psychosomatic conditions such as railway spine, neurasthenia, shell shock, and post-traumatic stress disorder. They are syndromes dredged from the symptom pool of the times.

It's hard to remember that courage, pluck, and gallantry are also part of the soldier's tale when so much of what we hear of the military after a war is drowned out by the gripes of the goldbrick or the sad sack. I spent a good bit of my time in military service performing discharge physicals on soldiers as they were mustered out after the Korean War; those who had fought the hardest had the fewest functional complaints. Neither noncombatant Commander Cop, the peacekeeper, nor engineer Kevin Rudland remind one of Keith Douglas's tank commander in the Tunisian desert of 1943:

> Peter was unfortunately killed by an 88
> it took his leg away, he died in an ambulance.
> I saw him crawling in the sand; he said
> It's most unfair, they've shot my foot off.
> How can I live among this gentle
> obsolescent breed of heroes and not weep?
> Unicorns, almost....
> The plains were their cricket pitch
> and in the mountains the tremendous drop fences
> brought down some of the runners. Here then
> under the stones and earth they dispose themselves,
> I think with their famous unconcern.
> It is not gunfire I hear but a hunting horn.[29]

December 22, 2000

Rock of Ages:
Why We Creak

MOST PATIENTS with osteoarthritis—elderly or youngerly—complain that they creak and want to know why. This week, jogging around the Jardin des Plantes in the December sunshine of Paris, I creaked most awfully, and when I limped back to our pied-à-terre on the Quai I was asked by our waiflike concierge why I kept hurting myself every morning. I told her that I've been running around like that in the morning for almost thirty years. "I didn't know that old men in America did that sort of thing," she huffed. In halting French I tried to rephrase a response given by Bruce Bliven to the question of how it felt to be an old man. "I don't feel like an old man," replied Bliven, an editor of the *New Republic*. "I feel like a young man who has something wrong with him." My French wasn't up to the task.

Subdued, I hobbled to my desk to review two new books on aging that tell us pretty much everything one needs to know about the subject. Steven Austad's *Why We Age* and Tom Kirkwood's *The Time of Our Lives: The Science of Human Aging* hit their marks squarely: both are briskly written, their arguments clearly presented, and the authors of each have larded their stew of good science with enough narrative, personal anecdote, and controversy to recommend them to doctor and patient alike.[1,2] While these books differ in their emphases, each has its own twist on what is known, or not known, about aging today. It is not surprising (and somewhat endearing) to find that the authors—both active workers in the field of aging—are better at discarding competing notions of aging than at convincing the reader of their own pet theories.

SHEDDING SOMA, BURNING OUT

Tom Kirkwood, as might be expected from a man who has formulated the "disposable soma" hypothesis of aging, begins by dispelling two of the most common ideas about aging. "The first of these is that ageing [sic] is inevitable because we just have to wear out. The second is that ageing is necessary and we are programmed to die because otherwise the world would be too overcrowded." Kirkwood also argues that, while the germ plasm (sperm and egg) may be continuous, the rest of our somatic cells (the soma) are a kind of chaperon of the germ plasm, and that like chaperones at a party, they are meant to look busy and disappear. But it takes energy to look busy and they wear out. Since energy requirements, and the various pathways of oxygen metabolism can be approached by the methods of modern science, we can solve this aspect of the problem. We don't have to wear out; indeed, some species have partially solved this problem already. The second erroneous notion—the overcrowding hypothesis and its corollary, the death gene—has all the baggage of teleology behind it and is neatly disposed of by the argument that neither the germ line nor somatic genes code for the kinetics of crowding.

Steven N. Austad dismisses two other notions of aging: that organisms die because their cells do not divide, and that our lives are finite because our cells contain preset fuses for cell death (read: telomeres). Austad shows a pretty sound distrust for in vitro work, as might be expected from a scientist who has studied the opossum in the wild and who is now studying the budgerigar, a species of bird with exceptional resistance to oxidative damage.

Austad has been taken to task for this approach by Leonard Hayflick—a solon in the field of aging—in a review uncharitably titled "Why biogerontologists should not write popular books on aging."[3] Hayflick rightly criticizes Austad for knocking "greedy reductionism" as an approach to the study of aging. Further, Hayflick appeals to readers "to judge for themselves whether the reductionist study of cells and their components in this century has contributed anything useful to our knowledge of genetics, development, immunology, pathology, aging, or any other area of biological or biomedical science."

Be that as it may, Austad makes a pretty good case for the usefulness of the Denham Harman free-radical theory of aging: we age because oxygen-derived free radicals wear us down. That's reductionist enough for anyone.

GIBRALTAR MAY TUMBLE

That same holistic/reductionist dialogue seems to dominate the community of those who study aging for a living. It also seems to affect those of us who have come to look at the problem late in our investigative careers. Holism, as Lewis Thomas informs us in one of his later essays, is a term "invented out of whole cloth by Gen. Jan Smuts in 1926"; it implies that matter and life are one.[4] Reductionism, on the other hand, derives from Hypolite Taine (*De l'Intelligence*, 1871), a positivist who borrowed the term from the chemists who use it to denote an agent that reduces a compound to a simpler substance by removing oxygen. Matter without life, one might say. Thomas was convinced that the problem of aging could readily be split into two definable subsets. The first was the sum of all the things that went wrong over time with the joints, the hormones, the arteries, the everything. But behind these phenomena lurked a deeper problem. Often obscured by individual pathologies—and therefore much harder to tackle—was what Thomas called "normal aging."[5]

Thomas was convinced that once the individual pathologies were taken care of, aging would still be aging and "a strange process posing problems for every human being."[6] He hoped that doctors would regard this process through lenses less reductionist and more holistic, regretting that the term had fallen on bad times with scientists. He should have been alive to read Austad's book. He would have loved the science, been tolerant of his prose, and admired Austad's organismic view, as opposed to Hayflick's reductionist tone.

As for myself, I'm not sure. I don't like the neo-Darwinian tone of the "aging genes" that have been postulated in worms or fruit flies; these seem to code for enzymes that permit cells and organisms to cope with oxygen and the free radicals derived therefrom. What neither Kirkwood nor Austad discuss is the simple truth that aging is not confined to the animal kingdom. Although inspired by Wallace and his description of the mineral world, Darwinian evolution does not apply to rocks. Rocks age. They age in part by oxidation, in part by hydrolysis, and in part by radiant energy; each of these reactions produces more or less unstable intermediates. We also age by the same inexorable mechanisms; all that enzymes do is to hasten reactions that take place in inanimate objects and fluids. What Austad calls "browning," due to glycation of macromolecules (adding sugar molecules to things like tendons, arteries, etc., as in diabetes), is simply an analogue of processes undergone by dead cellulose; if you don't believe me, look at your garden furniture.

Photographs fade, the pyramids may crumble, Gibraltar may tumble—all without the intervention of Darwin's genes or Hayflick's cell-doubling. Not telomeres, not prions, not SOD, not mitochondrial DNA—none of these has any role in the aging of rocks, nor for that matter, in the Rock of Ages. I would argue against the holistic approach and go straight for old-fashioned reductionism, down to rock bottom. My guess is that the underlying culprit is chronos and free radicals: time and oxygen wait for no man.

December 12, 2000

The Genome's in the Mail

THIS WEEK the vast right-wing conspiracy stopped an ongoing recount of presidential votes in Florida. The process was anticipated by Ambrose Bierce in his 1913 *The Devil's Dictionary*: "RECOUNT, *n*. In American politics, another throw of the dice, accorded to the player against whom they are loaded."[1] In other news, Arabs and Israelis battled in Bethlehem and Craig Venter of Celera announced that his company has submitted the complete sequence of the human genome for publication in *Science*. Celera anticipates publication of the manuscript in early 2001 and "continues talks with the International Public Human Sequencing Consortium for simultaneous publication."[2] Those continued talks were no less contentious than Scalia *v.* Boies in the Supreme Court and no closer to conclusion than the peace process in the Middle East. Pleading commercial constraints, Celera had attached strings to publication of its data, conditions that the editors of *Science* accepted but which the public consortium and its friends found unseemly.

Some gene hunters were up in arms after the editors of *Science* circulated details of the agreement with Celera; others found it unusual, but fair.[3] Under the terms of publication, academic users would have almost unlimited access to the database. They could perform searches ad lib and download segments of DNA up to one megabase (one million bases or digits of DNA); if they needed larger downloads, they would have to pledge not to redistribute the data. They could publish the results of their work and patent any novel observations without interference by Celera. On the other hand, commercial users would have to guarantee not to commercialize the results, nor could they redistribute any sequences used. "Alternatively," explained *Science*, "they may obtain a license or a subscription for a fee." *Science* will also keep a copy of the database in escrow, "to insure that there will be no changes in the ability of the public to have full access to the data results." Miffed that the data would not be deposited in the NIH GenBank—the customary archive for DNA sequences—David Baltimore of Cal Tech argued that "if the need for commercialization is so important, then don't publish. Don't get the accolades of publication."[4] Dr. Samuel Broder of Celera defended the agreement reached with *Science*, arguing that companies using the data should bear their fair share of research and development costs: "The bottom line is we want to make sure that the work and considerable effort is put to the benefit of the people who took the risk to invest in Celera."[5] Having become a CNN junkie by the Florida recount, I'd rule for Celera by a vote of four to three.

BLACK TIE AND TALES

Three days before, Venter, in black tie and high spirits, had received another sort of accolade. News that the manuscript was in the mail was first made public at the Annual New York Gala of the American Committee for Israel's Weizmann Institute of Science (Rehovot).[6] Venter received the Committee's Award for Excellence in Research for spelling out the human genome faster and for less money than anyone had thought possible. Venter assured his glittering audience of Nobelists, diplomats, and philanthropists that deciphering our genetic scroll was not only a milestone in medical science, but also "the foundation of pharmacology in the new millennium."[7]

Venter's citation by the Weizmann Institute also praised him for having developed ESTs (expressed sequence tags), a technique for picking out the real genes that constitute only 2 to 4 percent of our DNA from the vast silent majority of what is called "junk" DNA. He told his listeners that Celera had "combined the EST technique with the whole genome shotgun method" to blast out chunk after chunk of DNA and then to set each back in its proper order. Using this method he was able to map similarities and differences among the genomes of five ethnically diverse humans. Those differences are coded by "single nucleotide polymorphisms" or SNPs, pronounced "snips." Snips are found roughly once every 1,250 base pairs and account for over 90 percent of genetic diversity.[8] At last count, Celera had identified 2.4 million SNPs as against eight hundred thousand for its publicly funded rivals.[9] Venter ascribed his own success entirely to the whole genome shotgun technique because it gave him "uniform SNP coverage of the genome and the fact that every SNP could be assigned its exact genome location."[10] Venter was sure that by combining information of genetic variation from Celera's diverse donor pool (the snips) with the annotated human genome (the data coming out in *Science*), gene hunters will be able to predict variations in our response to pharmaceuticals, microbes, and trauma; drugs, bugs, and thugs, as it were.

Other benefits are likely to emerge from the genome project. As of today, no one really knows exactly how many genes there are in the human genome; estimates range from 28,000 to 150,000.[11] Reliable sources have it that Venter's data will show that the actual number is close to the lower estimate. Moreover, no one has really provided a generally accepted biochemical definition of a gene. Most definitions are indirect, based on the gene's dictation of RNA and protein synthesis, or entirely functional as "the structural unit of inheritance in living organisms."[12] To circumvent these difficulties, Lubert Stryer and I came up with a foolproof definition a few years ago at a meeting dominated by reports of gene knockout experiments. We recalled Otto Löwy's definition of a drug: "A drug is a substance, which, when injected into an animal produces a paper," and concluded that the biochemical definition of a gene must be "a unit of DNA, which, when deleted in a mouse, produces a paper."[13] By that criterion, Venter's data should launch a minimum of twenty-eight thousand papers.

NOT IN OUR GENES

Like Newton explaining the clock of the planets, Venter reported a grand conclusion he'd drawn from his map of the genome. There are 3.5 billion base pairs of DNA—coincidentally, one pair for every year of life on this planet. Differences between one or another of us constitute less than one-tenth of one percent of all those base pairs, and in each of the genomes he examined, the bulk of DNA sequences tallied up with one another and were independent of race. The genome, he concluded, gave proof of the brotherhood of man. That egalitarian conclusion sat well with his audience of Weizmann supporters. It was very much in keeping with earlier studies based on incomplete protein and DNA data. Richard Lewontin of Harvard and Luca Cavalli-Sforza of Stanford have argued for many years that many human traits—aggression, reason, race—are not in our genes. If we have trouble defining a gene, we face an impossible task with race.[14] Cavalli-Sforza has neatly documented that skin and hair and face—the outer marks of race—do not define who we are: "It is because they are external that these racial differences strike us so forcibly, and we automatically assume that differences of similar magnitude exist below the surface, in the rest of our genetic makeup. This is simply not so: the remainder of our genetic makeup hardly differs at all."[15] Molecular anthropology is correct, Venter reported in New York: we're brothers under the skin because we're linked by codons in common.

THE YS HAVE IT

On the platform of the Weizmann dinner, on a day that gunfire enveloped Rachel's tomb in Bethlehem, Venter was a reminder that the lessons of science for human fraternity have yet to sink in. Earlier this year in Riyadh, Saudi Arabia, Venter had been awarded the King Faisal Prize for biology.[16] I'd venture to say that no one in the two admiring groups Venter addressed this year—not in Riyadh, not in Manhattan—could explain why mistrust in the Middle East is so intractable, why the closest of kin wage the saddest of feuds. He won't find the answer in our genes, certainly not in one of those 1,250 snips that govern human diversity. Indeed, population geneticists from Arizona and NYU have just reported that the genomes of the two warring factions in Bethlehem are almost superimposable.[17] They based their conclusion on a neglected bit of human architecture, the Y chromosome of man.

Haplotyping the Y chromosome is becoming the gold standard of population geneticists who have learned that the tiny human Y chromosome is more than "a heap of evolutionary debris, hooked up to a sequence that happens to endow its bearer with testes."[18] That's because two men with the same genetic tags on their Y chromosomes must be descendants of the same male ancestor. The chromosomes of our somatic cells exist in pairs and therefore they can recombine when sperm cells are formed. That's why a son who has inherited a given chunk of DNA from *his* father's somatic cells may not necessarily pass it along to his son. But the Y chromosome doesn't recombine, because it doesn't have a partner. Like last names, it is transmitted intact, give or take a spelling error, from father to son for generations.[19]

Hammer et al. showed that the allelic variations—spelling errors—in the Y chromosome of Arab and Jew cosegregated from other groups.[20] The paternal genomes of Jews—whether they came from Europe, North Africa, or the Middle East—derived from a common Middle Eastern ancestral population. After the first major Jewish exile of 586 B.C., when Jews dispersed across Europe and North Africa, Jews largely retained their genetic identity, one that was formed in the Middle East. Even after centuries of exile, Diaspora Jews remained closer to each other and more similar to Palestinians, Syrians, and Lebanese, as judged by their shared Y chromosome tags, than to non-Jews of

Western Europe.[21] The factions battling over Rachel's Tomb are the closest of brothers.

A pessimist might argue that the mark of Cain is on the Y. I don't really believe that. Riyadh and Rehovot may be separated by custom, culture, history, and religion; powerful forces indeed, but not dictated by the genome. The lesson I take from the new genetics is that genes are innocent bystanders in the quarrel of peoples. Strung out one after another in the ranks of our genome, they're like peaceful ships of the line on the Mediterranean with their sailors ashore, as in Auden's 1955 poem *Fleet Visit*:

> But the ships on the dazzling blue
> Of the harbour actually gain
> From having nothing to do;
> Without a human will
> To tell them who to kill
> Their structures are humane.[22]

November 27, 2000

The Great Fear:

Mad Cows and Englishmen

THE NEWS this week carried overtones of the eighteenth century. In the United States we were preoccupied with a national election as undecided as that of 1800 when Aaron Burr and Thomas Jefferson received an equal number of electoral votes for president. In France, the nation was seized by a replay of the Great Fear of 1789; this time around the threat was not of foreign invasion and counterrevolt, but of mad cows and Englishmen.

THE GREAT FEAR

This week France was in panic over *la vache folle*—mad cow—disease. The country of *steak/frites* was off its feed and left with a million tons of bone meal to burn.[1] Cuts of tainted beef from a herd in which a case of mad cow disease (bovine spongiform encephalopathy, or BSE) had been diagnosed were found on supermarket shelves. Newspaper headlines warned that over 103 cases of BSE had been diagnosed in French cattle this year—versus 31 cases the year before.[2] Having banned beef imports from Britain, which underwent an epidemic of BSE from 1985 to 1990 with thirty-seven thousand cattle affected, the French had claimed superiority not only in culinary, but also sanitary matters. In Britain, eighty humans fell victim to a cross-species jump of the bovine brain disease. This variant of Creutzfeldt-Jakob disease (vCJD) differs both clinically and epidemiologically from the classical form, affecting younger patients with great severity. Due to its long latency, anywhere from five to fifteen years, the toll of vCJD in Britain is still rising.[3] La Belle France remained untouched; English but not Gallic beef was suspect. Until this year.

The Great Fear struck France after the second case of French vCJD was documented late this summer.[4] It intensified after beef from a *vache folle* was found on the shelf of the supermarket chain Carrefour. Television raised the stakes when it showed gruesome footage of a youthful victim of vCJD. It was the first time that the French public had seen a French victim of the disease. The boy, nineteen, once a fit athlete, had turned to skin and bones, his limbs were contracted at odd angles, his gaze vacant. The boy was being cared for at home; his father had to carry him about the house in order to feed and bathe him.[5] Beef consumption dropped over 40 percent in France, the price of beef fell over Europe, and newspapers grew shriller by the day. Supermarkets lost customers to *bouchers sérieux*, their trusted neighborhood butchers. Worse yet—the government announced that for one year it would ban sale of that most prized of organ meats, the treasure of French vianderie: *ris-de-veau* (calf thymus).[6] Panic hit agricultural and grain markets, rumors and broadsides filled the streets, menus were rearranged in bistro and three-star restaurant alike. A popular food alarmist, Dr. Saldmann, said that public officials had sent out mixed messages. "First, they said the beef was safe. Then they instituted new safety measures. So which time were they telling the truth?" Dr. Saldmann, asked. "Everywhere you turn right now, people are saying, 'Who can we believe?' "[7] Whispers and innuendo, mutterings against perfidious Albion and know-nothing Germania filled the editorial pages; the Great Fear was abroad.[8] The English problem had crossed the channel, the cordon sanitaire (the quarantine line) had been breached.

In France this week, rumor ruled the roost to an extent unknown since the episode of the Great Fear of 1789. In the confused weeks after the Bastille fell (July 14–August 4) rumor had it that France's European neighbors were taking advantage of her internal disorder. The Austrians and Prussians were invading French soil. They were stealing her grain and ruining her commerce. Counterrevolution was abroad. In response, châteaux were burned, shops were looted, Alsatian Jews pursued, the bourgeoisie humiliated, and innocents slaughtered. The peasants were aroused and the countryside aflame. Historians trace the greater terrors of 1792 to these first outbreaks of unreason: "Fear had broken out everywhere and at once, spread by mysterious messengers and engendering agrarian revolt."[9] Compared to the Great Fear, the *vache folle* crisis is small potatoes indeed. The whole affair conforms to Karl Marx's put-down of the Second Empire: history does repeat itself, but the first time as tragedy, the second as farce. This week in France, Michelet was rewritten by Feydeau: there were no tumbrels rolling to the guillotine, but the Concordes were grounded, the EU disdainful, and *ris-de-veau, abandonée*.

PROUST TO THE RESCUE

The government struck back. Prime Minister Lionel Jospin announced that he had appointed Jean-Paul Proust (a former prefect of the Marseille region and no known relationship to Marcel) to oversee what the quasi-official newspaper *Le Monde* called "Le plan de bataille contre la vache folle" (The battle plan against mad cow disease).[10] In the Cartesian fashion of French politics, which combines cool analysis with stubborn self-interest, the plan was based on four biological facts and an overriding economic imperative.

- The infective agent of BSE is an abnormal "prion" protein (PrPsc) that is spread by using bonemeal as livestock feed for hogs, chicken, and cattle. Cows, but not chickens or hogs, become infected when they ingest PrPsc. Bovine meat can then infect a small minority of humans who will come down with vCJD after eating PrPsc-contaminated meat.

- Bonemeal is the residue of beef slaughter (i.e., residues of animal carcasses, mainly bone and its marrow, once the meat and commercially valuable products such as tallow or gelatin have been extracted). Outbreaks of BSE almost precisely track exports of feed from Britain in the 1980s and early 1990s.[11]
- Cattle under thirty months of age *almost never*, and cattle under eighteen months of age *never*, carry the abnormal, infective prion. Instead they carry its normal counterpart called PrPc.
- Normal humans also carry that normal prion, PrPc, as part of the surface membrane of their brain cells. But, if things go bad for us and we encounter PrPsc from an infected cow, the incoming prion interacts with our normal prions to form insoluble, peanut brittle-like tangles that lead to a progressive, lethal disease of the central nervous system.[12]

Now for the economic imperative: if France bans all production and export of bonemeal, not only will it lose the economic benefit of producing the meal, but it will also bear the cost of burning it up and of growing or importing replacements such as soybeans—mainly from the U.S. According to Proust, this would cost a minimum of $650 million per annum.[13] As a result, French beef would become uneconomic to export.

Proust's courageous counterattack was to ban the production, import, and use of bonemeal in France; to test cows at risk; to track beef from areas in which BSE had been identified; and finally to ban use of brains, spinal cord, and, *hélas, ris-de-veau*. Left with the problem of discarding all that leftover bonemeal, he found refuge in the solid Gallic tradition of *defense de gaspiller* (never waste anything). The French are turning bonemeal into industrial fuel: "Meat-and-bone meal does not make very good fuel, but it's a fuel," said a manager at a cement factory owned by the international giant Lafarge.[14]

Jean-Paul Proust's defense of the culinary barricades may remind some of another sanitarian, Marcel Proust's father, Adrien Proust (1834–1903), who was professor of hygiene in the Faculty of Medicine at Paris. He was also founder of l'Office international d'hygiène publique, the precursor of the World Health Organization, and in that role he directed the *cordon sanitaire* of 1892 that quarantined France from the cholera epidemic itching to cross her borders from Spain and Italy.[15]

PRIONS WITHOUT BORDERS

France tried to persuade its sister republics to join in the battle against mad cow disease, and failed miserably. Last Tuesday, agriculture ministers of the European Union, meeting in Brussels, said that they would not join the French in banning use of bonemeal for feeding livestock in the EU. They grudgingly agreed to begin testing all animals "at risk" as of January 1, 2001. Unfortunately, "testing animals at risk" means performing slaughterhouse exams of old cows with "staggers" or those that simply keeled over on their own. There is no blood test available that can be used in live animals. Bad news for France, since Italy, Greece, and other nations would now be free to invade markets France has been forced to abandon. Of course, banning bonemeal over the fifteen-nation EU would have required destruction of 3 million tons of bonemeal every year and finding its replacement. No wonder that France's neighbors simply voted to ban French bonemeal and to halt imports of French beef; what was looming, of course, was a cordon sanitaire in reverse.

Then suddenly over the weekend two cows in Germany, one in the Azores, and one in Spain were found to have contracted the disease.[16] The German government was forced to do a U-turn and to demand immediate tests on all cattle over thirty months slaughtered throughout the EU. "The spread of BSE to Germany shows that BSE knows no national frontiers," European Health Commissioner David Byrne told German television on Sunday. Following rumors that perhaps dozens more of the Germany's 15 million cows were infected, butchers reported plunging beef sales—as in France—and the media howled for ministerial resignations. The Germans blamed their bovine misfortune on French imports and the French chalked up their disease to British meal.

Prime Minister Tony Blair resisted a ban on French beef despite widespread resentment by his fellow Brits that France refused to import the British product years after the rest of the EU ended its ban on the roast beef of Old England. The *Independent* sent a semi-apology: "Britain is badly placed to lecture others. The European B.S.E. epidemic is an extension of the British epidemic, spread largely by the morally unforgivable, even if legal, dumping of scores of thousands of tons of suspect animal feed on the Continent after it was banned in the U.K. in 1988."[17]

To calm the clouds gathering over Europe, French officials took out full-page advertisements in newspapers proclaiming the overall safety of French beef: "Why beef can be eaten without fear," the ads explained.[18] Nevertheless, the government went full speed ahead with testing—but only for cattle over thirty months of age who were at risk: old cows with staggers, again. M. Jean Glavany, the minister of agriculture, said that France was being punished for finding more BSE cases than Germany or Spain, only because Gallic experts were better pathologists, finding microscopic disease even in "partially infected" animals: French vets were counting "possibles" as "positives."[19] The pregnant chad principle, so to speak.

NOBEL TO VONNEGUT

The story of BSE, prions, and vCJD is a fine tale of pluck and persistence. The best version was written by Stanley Prusiner, himself, after receiving the Nobel Prize in 1997 for his discovery of prions.[20] He had found that the infectious agent of scrapie, a "slow virus" of sheep, so-called because of its long incubation period, could be propagated in hamsters. When he purified the agent more and more, he found not a trace of nucleic acids—as would be the rule in conventional bacteria, viruses, or fungi. "As the data for a protein and the absence of a nucleic acid in the scrapie agent accumulated, I grew more confident that my findings were not artifacts."[21] He summarized that work in a *Science* article published in the spring of 1982.[22] "Publication of this manuscript, in which I introduced the term 'prion', set off a firestorm," Prusiner wrote in his Nobel memoir. It sure did, and it took over fifteen years for the work to sink in. "Because the novel properties of the scrapie agent distinguish it from viruses, plasmids, and viroids," Prusiner wrote in *Science*, "a new term 'prion' is proposed to denote a small proteinaceous infectious particle which is resistant to inactivation by most procedures that modify nucleic acids. Knowledge of the scrapie agent structure may have significance for understanding the causes of several degenerative diseases."[23] Several, indeed.

Without exception, all known prion diseases lead to the death of those affected. There are, however, great variations in presymptomatic incubation times and how aggressively the disease progresses. The prion protein, designated PrP, can fold into two distinct conformations, one normal (PrP = PrPc) and another that results in disease (scrapie PrP = PrPSc). The disease-causing prion protein has infectious properties and can initiate a chain reaction. Final proof that the PrPc to PrPsc chain reaction is critical to the prion disease was afforded in 1993 by Charles Weissmann's lab in Zurich.[24] When the Swiss workers deleted the PrPc gene in mice, the knockout mice grew up normally. But, when these mice were inoculated with mouse scrapie prions, they remained free of scrapie symptoms for at least thirteen months while wild-type controls all died within six months.

The normal PrPc protein is a monomeric, signal-transducing, cell surface protein whose biological function is unclear. It is expressed mainly in nerve cells and only in neurons that are fully differentiated and that display receptors for neurotransmitters.[25] When mutated, as in genetic forms of CJD, or infected by PrPsc from diseased cows, the monomers of PrPc are converted to insoluble, disease-provoking meshworks of PrPsc. Accumulated over time and space, the aggregated polymers induce scrapie in sheep, BSE in cows, and vCJD in humans.[26]

Aggregation is spontaneous, depends on specific amino acid sequences, and is completely and utterly irreversible. Whether afflicted by genetic errors of PrPc, by injections of human pituitary growth hormone, or by *ris-de-veau*, many different kinds of people can succumb to self-aggregating prions. I've often compared that chain reaction to the sci-fi terror of Ice Nine in Kurt Vonnegut's *Cats Cradle*. Ice Nine is a structure of water that on contact freezes other water into an ocean of ice at room temperature; PrPsc is a device for turning cell sap into rock crystal. Only Vonnegut's Bokonon would think it:

> Nice, nice, very nice;
> Nice, nice, very nice—
> So many different people
> In the same device.[27]

There is perhaps a more scholarly appreciation of what the prion diseases are about. I first heard about scrapie a few years before the word "prion" ever appeared in print. Lewis Thomas was always persuaded that chronic autoimmune diseases such as rheumatoid arthritis or lupus erythematosus might well be related to the kind of slow viruses that cause scrapie. And, sure enough, he urged his younger associates to follow the scrapie literature, pointing out similarities between autoimmunity and scrapie: the slow onset, relentless progress, and our failure to find a conventional microbe at their root. In 1981 Thomas was on sabbatical in England and on his way to a conference devoted to scrapie. He had heard that Prusiner and his gang were on the spoor of the infective particle, and that it might be a protein. But how could a simple protein cause disease?

> Halfway along on the London streets it occurred to me that maybe the virus of scrapie is simply a switched-on, normal gene, and the presumably protein agent that caused the disease may not be alive after all, only a signal to switch on a gene in brain cells which is supposed to be kept off.[28]

Thomas got scrapie right in theory, just before Prusiner and Weissmann pulled off the much harder job of doing the experiments. They proved—irrefutably—that the gene lurks in all of us all of the time, simply waiting for the signal to become: Ice Nine. Nice, nice, very nice.

November 13, 2000

Pesticides: The Nader Factor
and Heidegger

THIS WAS the week that the presidential election became interminable and movement disorders were traced to their roots. In New Orleans, neuroscientists announced that rotenone, an eco-friendly pesticide extracted from tropical roots, might be responsible for Parkinson's disease. In Florida, the "Nader effect" reminded us that political movements based on technophobia have had dismal consequences in the twentieth century.

In a study that made headlines on election eve Dr. J. Timothy Greenamyre, a professor of neurology and pharmacology at Emory, reported to the Society for Neuroscience in New Orleans that he and his team had induced Parkinson's disease (PD) in rats by infusing rotenone for one to five weeks directly into their brains.[1] The rats had come down with the Full Monty of PD: muscle rigidity, tremor of the distal extremities, and bradykinesia. Under the microscope, their brains were stamped by the hallmark of the disease, intracellular deposits of nucleoproteins synuclein and ubiquitin, called Lewy bodies. In the substantia nigra region, the vortex of PD in the brain, the Emory group found "a progressive degeneration of the dopamine system that goes awry in Parkinson's disease." The work was immediately hailed at the meeting by Dr. John Q. Trojanowski of the University of Pennsylvania, who told reporters that, "This is the best model we have ever had for this disease being associated with an environmental agent." Greenamyre ventured that pesticides such as rotenone, which he said were "used in a zillion products," might indeed be responsible for the condition in humans. "Pesticides are essential for growing crops, but we may need to think about minimizing their environmental impact."[2]

While Greenamyer's complete data remain to be published in the December issue of *Nature Neuroscience*, much of what was presented in New Orleans had already seen light of day last year in an unrefereed symposium. Greenamyer claimed that "We [have] developed a novel model of PD in which chronic, systemic infusion of rotenone, a complex 1 [mitochondrial] inhibitor, selectively kills dopaminergic nerve terminals and causes retrograde degeneration of substantia nigra neurons over a period of months. The distribution of dopaminergic pathology replicates that seen in PD, and the slow time course of neurodegeneration mimics PD more accurately than current models."[3]

That little prepublication conflict remains to be explored, but the full story of rotenone, mitochondrial complex 1, and Parkinson's disease is a tribute not only to pharmacology, but ethnobotany, epidemiology, and simple clinical observation. A single case report in 1979 begins the tale.[4] A twenty-three-year-old man died after chronically abusing a meperidine (demerol) analogue and developing severe, unremitting Parkinsonism. I. J. Kopin and his NIH colleagues recognized that, unlike other drug-induced motor disturbances that have acute outcomes, the young man's syndrome persisted for eighteen months. After biochemical and pathologic studies suggested that the illicit drug had decimated dopamine-producing neurons, the patient responded partially to drugs that stimulated dopamine receptors. As I note (October 17 2000), Arvid Carlson of Sweden was awarded this year's Nobel Prize for physiology and medicine for showing that one can produce Parkinsonism when dopamine in the substantia nigra is depleted by means of reserpine and that giving dopamine precursors can permit Awakenings à la Oliver Sacks. The NIH group correctly concluded that the meperidine derivative had punched selective holes in the young man's substantia nigra. One might say that he was a man who mistook his brain for a sieve. Kopin et al. were confirmed in a 1983 report of four more young addicts who developed severe PD-like symptoms after intravenous abuse of various meperidine congeners.[5] The chief chemical culprit was identified as 1-methyl-4-phenyl-1,2,5,6-tetrahydropyridine (MPTP) and the Parkinson research community was off to the pharmacologic races.

It soon turned out that MPTP, and other heroin analogues such as 1-methyl-4-phenylpyridinium (MPP+), are taken up by the dopamine transporter of dopinergic neurons where they inhibit the activity of mitochondrial complex 1 (technically, the NADH: ubiquinone oxidoreductase multienzyme complex, or EC 1.6.5.3). In animals given MPP+, energy stores drop in the brain and toxic free radicals (= *radix*, as in the L. for root) kill neurons in the substantia nigra. Stimulated by these lab findings, clinical investigators found that nonheroin-abusing patients with PD also had a profound defect of complex 1 in the mitochondria of their dopamine-rich cells, mimicking the effects of MPP+ in man and mouse. Consensus soon had it that PD was due to "a genetic or environmentally induced abnormality of mitochondrial function or free radical metabolism." Free radicals were the root of all evil in Parkinsonism.[6]

ALONG CAME ROTENONE

When the genetic load was found to be small, attention was turned to the environment. Large epidemiologic surveys showed a positive correlation of PD with tea drinking, a negative one with coffee; there was a sevenfold increase of PD in folks who worked or lived in the country as opposed to city dwellers.[7]

Pesticides seemed a likely source, but the data were conflicting. Then along came rotenone. Rotenone is an organic—that is, nonsynthetic—pesticide used widely on garden fruits and vegetables and for killing unwanted fish in lakes and rivers. Rotenone and its congeners are extracted from cube resin found in the roots (*radix*, again) of trees and vines with names like those of a Miami law firm: Stinkwood, Derris, Urucu, and Jewel Vine. From the jungles of Brazil, where native folk discovered their magic, to the Everglades of Florida, these plants have evolved rotenoids as homemade vermicides.[8] Unlike malathion or DDT, rotenoids are dissipated by air and water in a few days and are therefore greatly in favor with fans of the wild. Earlier this fall, at a rally in New York held to protest spraying against West Nile virus, Ralph Nader urged the city to turn from "synthetic poisons" like malathion to "organic controlants" obtained from plants and soil.

For at least 150 years, cube root and rotenoids have been used as fish poisons by tribesmen in the Amazon, pesticides by outbackers in Australia, and garden aids by the Martha Stewarts of suburbia. Often dusted onto roses, tomatoes, pears, apples, African violets, and household pets (!), it also kills fire ants. Over the last three decades, biochemists have worked out that the acute toxicity of rotenone to insects, fish, and mammals is attributable to—guess what—inhibition of mitochondrial NADH: ubiquinone oxidoreductase, our old friend complex 1. Due to this property, fishery managers can use rotenone in lakes and reservoirs; once it has killed unwanted fish by uncoupling their breathing machinery, the water is restocked with desirable species.[9] Rotenoids have other actions: they are promising candidates as anticancer agents since dietary rotenone reduces the incidence of liver and mammary tumors in rodents. Rotenoids are potent antagonists of tumor promoters in the dish.[10]

In New Orleans, Dr. Greenamyre presented the latest chapter in this story: rotenone, like the heroin/meperidine congeners, produced PD in rats and did so by inhibiting complex 1. He came close to claiming that he'd found a radical solution to the riddle of Parkinson's disease. Rotenone, or a similar environmental agent, he argued, evokes a Baroque flourish of ion fluxes (mediated by NMDA receptors) in the brains of Parkinsonian patients that ends in "free-radical production and mitochondrial depolarization." He was confident that, in this scenario, synthetic NMDA-receptor antagonists "may be neuroprotective."[11] Synthetic chemicals to reverse disease brought about by a "natural" product? Little did Dr. Greenamyre realize that by berating a "natural" plant extract—and one discovered by Amazonian tribesmen at that—he had struck a blow at the central dogma of the "deep ecology" movement.

GREEN ON TOP, BROWN ON THE BOTTOM

Last week brought proponents of deep ecology—the chlorophyll of the Green Party—to the surface of presidential politics in America. Deep ecologists and their fellow trekkers seem persuaded that every natural product pleases and only man is vile. Compare, for example, the new self-righteousness of the Green, as spelled out by Barbara Ehrenreich in *The Progressive*: "From the point of view of all other terrestrial life, we are the piggies of the planet—a hideous, death-dealing, global blight"[12] with the old self-righteousness of the Cross, as in Reginald Heber's *Hymn to a March*:

> What tho' the spicy breezes
> Blow soft o'er Ceylon's isle;
> Though every prospect pleases,
> And only man is vile?
> In vain with lavish kindness
> The gifts of God are strown;
> The heathen in his blindness
> Bows down to wood and stone.
>
> HEBER (1783–1826)

Another, somewhat less flattering pairing sprang to mind. Ms. Ehrenreich explained to readers of the *Nation* why she was voting for Ralph Nader: "What I fear most about a Gore victory—yes, I said victory, is its almost certainly debilitating effect on progressives and their organizations."[13] It seems to me I've heard that song before with less than attractive overtones. "So great was the sectarian divide in those crucial months before the deluge [1932] that the communists preferred even to link up and stage strikes with the fascists rather than campaign in the country and in the factories for a unified force against fascism. *'Nach Hitler Uns'* (After Hitler, our turn) cried communist leader Ernst Thälmann."[14]

Making his final pitch to an enthusiastic crowd on election eve, Ralph Nader misquoted the granddaddy of the Greens, German philosopher Martin Heidegger (1889–1976): "The basic fact about human beings is that we care about one another."[15] Actually, Heidegger was a bit more diffuse on the subject. What he actually wrote was that people "all belonged at once to each other,"[16] a notion derived from a long German tradition of kinship with the earth, *Blut und Boden* (blood and soil). In his major work *Sein und Zeit* (Being and Time, 1927), Heidegger anticipated Bill Clinton by asking what "is" is. He came up with the novel answer that "existence is the being of being there (*Dasein*)."[17] Sounds like an ad for a French perfume or a Japanese sedan.

But the new Greens and the eco-radicals among them are probably less attracted to notions of *Dasein* than to Heidegger's deep distrust of technology. They have found in *Die Frage nach der Technik* (The Question Concerning Technology, 1955) support for the belief that "technology is the supreme danger to man. The essence of technology, in the Heideggerian sense, is the supreme danger because it prevents us from having a proper understanding of our own being."[18] Heidegger's disdain for modern science, technology, and urban life persuaded him to spend his later years in a Black Forest retreat where he addressed villagers in a primitive Alemannic of his own device. Life in the forest and the radical ecology movement fit snugly "within the complex cultural and political terrain of the late twentieth century [in] relation to Martin Heidegger's antitechnological thought, 1960s counterculturalism, and contemporary theories of poststructuralism and postmodernity."[19]

When Germans despise something, they put it to question; they are, after all, a nation that invented the *Judenfrage* (the Jewish Question). Thus, when Heidegger posed "The Question Concerning Technology" his readers knew quite well that whatever it was, he was against it. On the other hand, they also knew what he was very much for: Adolf

Hitler and his brownshirts. Heidegger joined the Nazi Party in 1933 and in consequence was appointed rector of Albert-Ludwigs-University in Freiburg. His inaugural address declared his allegiance to Hitler: "Let not ideas and doctrines be your guide. The Führer is the only German reality and its law."[20] Ideas and doctrines were a guide to *some* of his behavior, nevertheless. He presided over the Germanification of his university by replacing Jewish professors with devotees of the Führer's blood and soil. Heidegger spent the Hitler decades writing impenetrable tomes that explored the difference between "earth" and "world" to arrive at such conclusion as "work lets the earth be an earth" or that man is the ruination of the natural order. Ruminations of this sort caused Lewis Thomas to moan that "The only sure memory I retain of Heidegger, even when I reread him today, is bewilderment."[21]

When the Green Party took the field in Germany after the war, their antitechnology, antiscience stance seemed a throwback to the *Blut und Boden* slogans of the Nazi brownshirts. Their opponents in the Social Democratic party soon sniffed the spoor of Heidegger and accused them of sharing his politics: like lawn turf, they were said to be "green on top and brown on the bottom." In the Nazi years, brown and green had united to overthrow the Weimar Republic. In postwar Germany, their common enemy was Germany's economic miracle. Their heroes were Heidegger and a packet of back-to-nature philosophers beginning with Jean-Jacques Rousseau: "Modern Green writers can rightly claim Rousseau as one of their most influential fore-fathers."[22] Rousseau's fancy of the noble savage and his general distrust of reason are now shared by green and brown the world over; the favorite targets of radical ecologists seem to be commonsense environmentalists like Al Gore.[23]

What about the gaunt Rousseau of our own day? Well, it's no surprise that Nader got his highest vote percentage in Alaska, where Bush and the oilmen also did well: loners and drillers, greens and browns. Ralph Nader has been bashing liberal democracy and its entrepreneurial arm since March 1960, when he wrote "Business Is Deserting America." In this earliest of publications, he warned of "our ingrained gullibility to internationalism." It was a warm-up exercise for his influential jeremiad *Unsafe at Any Speed* (1965), a book that helped shift the manufacture of gas-guzzling, unwieldy vehicles from American to Japanese venues. (It could be argued that thanks to Nader, we now drive dreadnought SUVs instead of snazzy Eldorados.) In keeping with Nader's recent concern for Iraq, it has not gone unnoticed that Nader published his 1960 piece in the *American Mercury*, a violently anti-Semitic magazine.[24] While he may have disowned those earlier companions, he hasn't changed his views of the corporate world. Nor have his supporters, who immediately after election day claimed that the 97,415 votes garnered by the Florida Greens were a step "to end the crisis of the environment and the corporate dominance of our government and culture." "We're on the map now," said Julia Aires, sixty, of Sarasota. "I think the goddess, in her infinite wisdom, put a pox on both their evil houses."[25]

The man who pegged this back-to-nature crew just right is Victor Klemperer, a cultural historian who was dismissed from his professorship in Dresden by a rector of Heidegger's persuasion. Here is his diary entry for Thursday, July 30, 1936: "Yesterday and today I worked through Rousseau's Encyclopedia article *Economie politique*; whole passages could be from Hitler's speeches. . . . The prostitution of reason in the service of subjective feeling, Romantic longing . . . the obsession with virtue as antidote and self-deception."[26] Obsessed with virtue as antidote and self-deception, the fans of Rousseau and Heidegger, no less than the eco-radicals around Ralph Nader, have a lot to answer for at the bar of reason.

October 30, 2000

Ebola: Out of Africa
with the Sanitarians

EBOLA HEMORRHAGIC FEVER

THIS WEEK the news came out of Africa where, at last count, Ebola hemorrhagic fever had killed 73 of 224 victims in the Gulu Province of northern Uganda. The Word Health Organization, whose frontline teams are monitoring the outbreak, warned that more deaths would follow over the next few months.[1] Ebola fever is essentially incurable, spread by human-to-human contact (skin, mucous membranes, blood) and kills by provoking massive bleeding. The virus replicates in endothelial cells and kills them: blood exudes from mouth, gut, and bladder, seeping through skin to form hemorrhagic blisters. The incubation period is between two and twenty-one days; doctors and nurses who care for the victims are particularly at risk.[2] Indeed, the latest victim was a nurse, and of the first four victims from whom the virus was isolated in Uganda, two were nurses and one a doctor. Experts from the CDC in Atlanta, whose teams are also on the site, have identified the strain as "Ebola Sudan." Journalists identify Sudan-based Ugandan rebels of the Lord's Resistance Army (read: thugs) as bearers of the epidemic.

The northern regions of Uganda, which the Nile crosses as it flows from Lake Victoria to the Sudan, are subject to constant incursions. The Ugandan Lord's Resistance Army (LRA) harasses the villagers by machete and Kalashnikov. The Gulu Province has been targeted by the rebels, who have kidnapped thousands of children, turning boys into child soldiers or weapon porters and girls into sex slaves. Uganda's long civil war between the LRA and the local villagers, the Acholi, has been waged since 1986 and continues to this day. David Westbrook, a student of the conflict, reports that "the death toll is unknown but probably is in the tens of thousands with many thousands more maimed and disabled. . . . Whether in protected villages or at home, the Acholi have been powerless to stop the LRA from burning their homes, schools, and clinics."[3]

Thanks to Ebola, the LRA is not the only army in the field. On the other side are platoons of sanitarians from WHO, the CDC, the Red Cross, and Médecins sans Frontiéres, who with gloves, masks, and sterile plasma are forced to travel in armored vehicles to fight the virus and heal the sick. After rebels abducted twenty-seven people in the area of the outbreak last Sunday, armed convoys were formed to protect doctors in the bush, according to Peter Odok, resident district commissioner from Gulu.[4]

Ebola first broke out in 1976 simultaneously in the Sudan and in Zaire, Uganda's neighbor to the west. Between June and November 1976 the Ebola virus infected 284 people in Sudan, with 117 deaths. In Zaire, which is now called the Democratic Republic of the Congo (but remains nondemocratic), there were 318 cases and 280 deaths in September and October. In 1995 an epidemic in Kikwit, Congo, killed 245 people. All in all, before this month's outbreak in Uganda, Ebola virus had killed more than 800 of nearly 1,100 documented cases.[5] The index case in Uganda was Esther Awete, a villager who died in her mud hut in Kabedo-Opongon on September 17 after only a day or so of fever and pain. Awete's chest began hurting. She became feverish and vomited blood. "We thought it was malaria," said Justin Okot, a neighbor. In the nearby town of Gulu, Awete was given chloroquine and sent home. "She didn't even last twenty-four hours," said Okot. "We didn't understand that someone could die that quickly." The villagers called the disease *gemo*—or evil spirit. "No one knows about it, but it comes and takes you in the night." Seven of her family members also died after they had ritually bathed Awete's corpse and washed their hands in a communal basin as a sign of communion with the dead. They then buried her thirty feet from where she had died.[6]

Ebola has not been the major cause of death in Uganda. Idi Amin (1971–79) and his hired *gemos* killed over three hundred thousand opponents; guerrilla wars and repression under Milton Obote (1980–85) took another hundred thousand lives. Uganda's President Museveni, chairman of today's governing National Resistance Movement (read: resistance to democracy), maintains that he heads a "movement that claims the loyalty of all Ugandans." The country harbors 205,000 Sudanese refugees who have fled south from two decades of civil war (black Christians and animists in the south against the Arab-Muslims of the north) and loss of 1.5 million lives to war and famine. Perhaps in response to such neighbors, Uganda's most recent "constitution" requires suspension of political activity until a referendum is held on the matter "sometime in 2000." Uganda's own civil war is expected to extend the deadline. Meanwhile, life expectancy in Uganda is forty-three.[7]

EBOLA IN THE LAB

If there was turmoil in the field, there was concord in the lab. Molecular biologists from Mount Sinai in New York and Marburg in Germany (speaking of tribal rapprochement) reported that they have worked out why the Ebola virus is so deadly and why we fail to rouse interferon against it.[8] Ebola viruses are enveloped, negative-strand RNA viruses belonging to the family Filoviridae that were first identified in 1989.[9] These viruses have small genomes of approximately nineteen kilobases that encode eight known proteins of which several, including VP35, VP40, a glycoprotein (GP), and an L-polymerase, have been identified as good targets for vaccine development. Gary Nabel's vaccine group at the NIH has shown that one of these proteins—a membrane-bound GP—is the most implacable foe of human endothelial cells. The virus, or the GP purified from the virus, not only kills individual cells, but breaks down connections between them. Gene transfer of GP into explanted human or porcine blood vessels causes massive endothelial cell loss that within hours leads to substantial increases of vascular permeability. When Nabel & Co. depleted a critical mucinlike region of GP, they abolished these effects—without affecting protein expression or function. Indeed, the GP from a strain of virus lethal to monkeys but not to humans (the Reston strain, named after the suburb in Virginia where it was first isolated) failed to disrupt human blood vessels while remaining deadly to monkey veins.[10]

Whereas Ebola uses GP as a lethal weapon against blood vessels, the virus has another artful tool on hand to direct its own packaging. The virus instructs infected human cells to form thousands of little matrices, or armatures, on which to stick its RNA. Last summer, scientists from the European Molecular Biology Laboratory in Grenoble found that monomers of the matrix protein VP40 order the assembly of that lethal package. Using chemical cross-linking studies, electron microscopy, and membrane-binding analyses, they discovered that the master switch to virus assembly was the N-terminal part of the VP40 monomer, without which the matrix cannot be properly formed to permit transmission of viral RNA.[11] And this week another piece of the Ebola puzzle was put in place. The Mount Sinai and Marburg scientists found that Ebola virus VP35 protein prevents us from making type I interferon, our time-honored means of reaching accord with a virus. The viral protein not only blocked RNA- and virus-mediated induction of an interferon-stimulated reporter gene, but also the induction of the interferon promoter. (We can't afford to buy the car, and if they gave it to us, we couldn't turn the ignition on.) The scientists concluded that since VP35 inhibits induction of type I interferon in Ebola virus-infected cells, it is likely that this mechanism underlies Ebola's virulence in Uganda.

THE TWO ARMIES

Deprived of internal accord, attacked by implacable foes, and subverted by self-made devices, the vascular system implodes when the Ebola virus strikes home. A cynic might argue that the history of postcolonial Uganda, from Idi Amin's *gemos* to the Lord's Resistance Army to the Movement of National Resistance has a parallel in the virulent, self-perpetuating scourge of Ebola hemorrhagic fever. However, not only in sub-Saharan Africa have dangerous thugs presided over famine and terror, children been sold into slavery, and civil wars left the stricken to die. Not only sub-Saharan Africa has required help from the sanitarian legions of WHO, the Red Cross, and Médecins sans Frontiéres. There've been two armies on many fields.

Only four years before his own young son was gravely wounded on the bloodiest day of civil war on this continent (Antietam, September 17, 1862), Dr. Oliver Wendell Holmes addressed the Massachusetts Medical Society to plead the sanitarian cause:

> As Life's unending column pours.
> Two marshalled hosts are seen,—
> Two armies on the trampled shores
> That Death flows black between
>
> One marches to the drum-beats roll,
> The wide-mouthed clarion's bray,
> And bears upon a crimson scroll
> "Our glory is to slay."
>
> . . . .
>
> Along [the other's] front no sabres shine.
> No blood-red pennons wave;
> Its banner bears the single line,
> "Our duty is to save."[12]

From Gulu to Grenoble, Marburg to Mount Sinai, no blood-red pennons wave, but the virulent may have met their match.

October 24, 2000

The Icelandic Genome Project
and Samuel Eliot Morison

THIS WEEK two major discoveries were reported by deCODE, a private genetic company in Iceland. DeCODE announced that its scientists had found a gene closely linked to schizophrenia. Simultaneously the company reported success in deciphering a critical gene that goes awry in lupus erythematosus. For commercial reasons, only one of those discoveries was made public. We can now start picking lupus apart, but schizophrenia remains a mystery.

FISH AND CHIPS

Last Friday, deCODE scientists together with psychiatrists from the Icelandic health care system proudly announced that they had identified "a gene linked to schizophrenia."[1] Based on a genome-wide screen of four hundred schizophrenia patients and an equal number of their unaffected family members, the study was so convincing to deCODE's partner in the gene-discovery business (Roche, of Basel) that deCODE received an "undisclosed milestone payment for its part of this accomplishment." Kari Stefansson, the former Harvard neurobiologist who returned to his native Iceland to found deCODE in 1996, noted that

"We are deeply grateful to the Icelandic patients who participated in this study. Their generosity . . . and of the physicians who support our efforts have moved us one step closer to understanding the genetic basis of schizophrenia." No further details were given, but one assumes that the genetic fact will stand the test of time—as well as the shrewd oversight of Swiss executives. The schizophrenia gene is only one of several genes deCODE has identified since 1998, including critical loci for osteoarthritis, Alzheimer's disease, and preeclampsia. The putative schizophrenia locus certainly sounds promising since its discovery has been based on the best of modern genetical science. Were the data available to scrutiny, one might judge how the finding might apply to other populations. But kudos are due, nevertheless.

Up to now genetic loci for schizophrenia have been elusive. Experts in the field confess that "Schizophrenia is a common chronic and disabling brain disease of unknown etiology, pathogenesis, and mechanism. . . . The inherited biological susceptibility to schizophrenia is probably expressed clinically as nonpsychotic abnormal personality traits, plus numerous biological markers (cognitive, anatomical, and psychophysiological) that are all found significantly more commonly in the population than is schizophrenia."[2] I interpret this as: "not a clue." After the announcement from deCODE the rest of us are still clueless.

Last Tuesday evening, as lights sparkled from the East River in New York, Kari Stefansson announced his recent discoveries to a select group of scholars and entrepreneurs. He vigorously defended deCODE's published policy of *not* making its results public immediately. Stefansson is a tall, intensely competitive intellectual whose devotion to the warrior/bards of Iceland is matched only by his passion for the poetry of William Carlos Williams. He argued cogently that the goal of biomedical science is to diagnose, prevent, and cure diseases, aims that often require that intellectual property be secured (read patented). By necessity and by law, protection of intellectual property will almost invariably lead to delays in publication. "The choice between early publication and the development of a product for the benefit of patients with a particular disease is, in our minds, an easy one."[3] I'm afraid he's right. No patent protection, no biotechnology.

One certainly hopes that deCODE has that locus right, that the discovery will help to unravel schizophrenia and that Roche got their money's worth. Iceland will clearly benefit, not only from reversing its chronic brain drain by bringing back scores of its native scientists. DeCODE's success in gene hunting is based on its access to a population genetically homogeneous, with founder effects for many traits, and on a well-documented genealogy of the entire nation that dates back to the first millennium. The Icelanders can trace all those Stefans to the son of the first Stefan among the original Vikings and Gaels who settled the island from A.D. 870 to 930.[4] "Icelanders have no grand monuments, no great palaces," Thordur Kristjansson (Kristjan's son), a genealogist, told a reporter. "The only thing we have had to remind us of our past is our family memory. Our genealogy is our monument." And sure enough, geneticists at the Icelandic Cancer Society were able to track arrival of a rogue gene in breast cancer, the BRCA2 gene, to a sixteenth-century cleric named Einar.[5]

For the last four years, the company has been encrypting a database with information on the health care of the entire nation, relating the island's genealogy (*The Book of Icelanders*) to genotypic information on a large part of the nation. It also has entered into an agreement with Affymetrix, Inc., to see if their microchips can be used to do genome-wide searches for known disease genes. DeCODE, which went public this July, also plans to use the database to search for drug targets in chronic diseases and to solve pharmacogenomic problems.[6] Its founders argue that basing the company in Iceland benefits the population directly by providing employment and returns on investment. After rigorous debate, the Icelandic parliament in January of 2000 granted deCODE the exclusive right to create and operate the health care database for twelve years, in return for $12 million. Iceland could earn up to an additional $1 million a year in shared profits and its health care system can use the database for free. The company's project is supported by close to 80 percent of Iceland's population, according to

Gallup polls. Most Icelandic political leaders agree: an executive of the national health ministry, Ragnheidur Haraldsdottir (Harald's daughter), notes that one of every thousand of Iceland's 273,000 inhabitants is now employed by deCODE. He anticipates that Iceland will become a world center for biotechnology, creating high-paying jobs in a country far too reliant on fishing as its major commercial asset.[7] Last Tuesday in New York a wag in Stefansson's rapt audience quipped that Iceland has gone from fish to chips.

In contrast to the hush-hush on the schizophrenia front, deCODE's breakthrough on lupus hit the press this week in a very visible immunology journal.[8] It's been known for decades that those missing one or another of the serum proteins of the complement system, females especially, are likely to come down with systemic lupus. Each of us requires the complete complement system (C_1 to C_9) not only to fight infection, but also to identify and clear debris formed when our own cells undergo programmed cell death in a process called "apoptosis."[9] Menstruating women deal with a wave of apoptosis every month. Most of the autoantigens of lupus (DNA, RNA, Ro, and La) that are usually found *within* cells become transiently displayed on their periphery in "apoptotic blebs" which form when the uterus sheds its lining.

The best-defined complement deficit is that of C_4 and genes that code for the two components of C_4, C_4A, and C_4B, are located within the major histocompatibility complex (MHC) on the short arm of chromosome 6.[10] The deCODE scientists knew that a large deletion covering most of the C_4A gene and the 21-hydroxylase-A pseudogene (on the extended haplotype B8-C_4AQo-C_4B1-DR3) probably accounts for two-thirds of C_4A deficiency in Caucasian systemic lupus erythematosus (SLE) patients. However, it's been tough to pick out the C_4A null patients due to the high degree of homology between C_4A and C_4B. The Icelanders devised a novel polymerase chain reaction (PCR) strategy to genotype for the C_4A deletion. They designed primers based on unique sequences upstream of C_4A as well as a human endogenous retrovirus that is present in C_4A but absent in lupus patients. From Reykjavik to Rome, Nome to New York, we can now begin to screen our SLE patients and their relatives by this means. A patent is surely in the works and deCODE should do well by doing good.

ELIZABETH I.COM

After Stefansson presented his arguments for deCODE's doing good, Professor Dorothy Nelkin of NYU, a distinguished scholar and sociologist of medicine, summarized many ethical objections to the project that have been raised in Iceland and elsewhere. She reduced these to three themes: the project invades privacy, it is too narrowly commercial in outlook, and deCODE can never obtain truly informed consent from all subsets of the population. Professor Nelkin echoed notes sounded by dissident Icelanders themselves: "Even aside from the privacy issue, which we feel is the critical issue, there is the question of why one private company is being allowed to take so much from Iceland," Tomas Zoega, psychiatrist and chairman of the ethics committee of the Icelandic Medical Association told the *Boston Globe*, while Jorunn Erla Eyfjord of the Icelandic Cancer Society added that "Kari is using a great talent for salesmanship to take over a country's medical records and exclude other scientists."[11] Petur Hauksson, chairman of the chief opposition group, Mannvernd, complained that "Iceland's healthcare information has been commercialized and our genetic information has been turned into a commodity."[12] Stefansson responded by arguing that the unique structure of Icelandic democracy, its national health service, its strong base in national altruism—and deCODE's unrelenting attention to ethical issues—will assure the philanthropic thrust of the company's research: "The reason we have medicine as it exists today is because our parents and grandparents supported and participated in medical research." Given a choice between abstract privacy and the fruits of research, Stefansson was certain that many—not only in Iceland—would opt for the latter. DeCODE's leader repeated his *NEJM* conclusion that "private enterprise has increased, not decreased, the pace of scientific discovery. Furthermore, the focus of the discovery of new knowledge (at least in biology and information science) has drifted away from academia and toward industry. Time alone will tell whether this is good or bad."[13] If you're an AIDS patient waiting for a new protease inhibitor from Vertex, for example, it looks good in the short run.[14]

Judging from history, private enterprise has been pretty good to discovery in the long run as well. Listening to Stefansson in a clubby venue as New York's East River flowed by, I was reminded of Samuel Eliot Morison, that great historian of our marine history, whose statue guards the greensward opposite St. Botolph's Club in Boston. His sagas of our settling remind us that we are the children of private enterprise. The Northeast of North America was first explored by Europeans of the privately financed Dutch West India Company (organized 1623). The seashores of Virginia were first the venture capital playground of the London Company and later of the Virginia Company (1607–9), while the Massachusetts Bay Colony (1628–29) promised to the pilgrim "Adventurers" (read: stockholders) "that the country would prove both beneficial to this Kingdom and profitable to the particular Adventurers."[15] Perhaps more to the point, exploration of the northern seas from Iceland to Greenland to Baffin and Frobisher Bays was financed by the Company of Cathay (1577). Chief stockholder of this privately financed quest for the northwest passage to China was Queen Elizabeth I of England. The Virgin Queen was in for £1,000, and a generous kitty of £4,277 was raised by a brace of others. The company funded all three voyages of the intrepid Adventurer Martin Frobisher (1535–1594), whose commercial success was celebrated in song and in Thomas Ellis's verse,

> The glittering fleece that he does bring in value sure is more
> Than Jason's was or Alcides fruite, wherof was made such store . . .
> And bringes home treasure to his land, and doth enrich the same,
> And courage gives to noble heartes, to seek for flight of fame[16]

Much of what we know of Frobisher, of the Cathay Company, and of discovery in the Elizabethan era dates to his logbooks: "The Three Voyages of Martin Frobisher" (London, 1938), edited by Vilhjalmur Stefansson (1879–1962). Stefansson, Stefan's son, would have been pleased that Kari, his Icelandic kinsman, "brings home treasure to his land, and doth enrich the same."

October 17, 2000

Nobel Prizes, the Mouse Genome Project,
and Aldous Huxley

MIXING MEMORY AND DESIRE

THIS WEEK the news came from Sweden and California. Stockholm
announced that Eric R. Kandel of Columbia, Paul Greengard of Rockefeller, and Arvid Carlsson of Gothenberg Universities had won the
2000 Nobel Prizes in physiology and/or medicine for their work on the
biochemical basis of learning, mood, and memory. From Pasadena
came news that we can fulfill an ancient desire: longer life without
decrepitude, at least in the lab. Shortly after a packed meeting at Cal
Tech was told which genes control the age of mice, Craig Venter
wowed the audience with news that he can now pinpoint those genes
precisely. His company, Celera, has mapped close to 95 percent of the
mouse genome in less than six months.

THE MEMORY LINGERS ON

The news from Stockholm would have made Sigmund Freud very
happy, indeed.[1] Kandel, Greengard, and Carlsson have largely realized
his hope that our higher mental faculties—such as mood, memory, and
desire—might be reduced to their chemical base: a series of reactions

controlled by phosphorous. The Nobelists had provided proof that one of Freud's heroes, Ludwig Büchner, was right when he announced the motto of scientific materialism: "Ohne Phosphor keine Gedanke" (no phosphorous, no thought).[2]

Carlsson found that dopamine was concentrated in the basal ganglia which control motor behavior. Using reserpine to deplete the brain's store of dopamine in mice, he found that the animals developed staggers and other abnormalities reminiscent of Parkinson's disease. Their symptoms were reversed by feeding L-dopa, a dopamine precursor. Why? Because when dopamine meets its receptors on nerve cells it starts a chain reaction that turns phosphorous into thought. Carlsson and others soon found that Parkinson patients have low concentrations of dopamine in their basal ganglia, and they were able to develop L-dopa as a drug to treat Parkinson's disease. The magic results are described in Oliver Sacks's small masterpiece, *Awakenings*. No Carlsson, no Robin Williams (who played Sacks in the film). The development of new anti-schizophrenia drugs depends largely on Arvid Carlsson's finding that Prozac affects synapses by blocking dopamine receptors.[3] No dopamine, not to speak of serotonin, no *Listening to Prozac*.

Eric Kandel and Paul Greengard's work intersects at the level of cyclic AMP, that ubiquitous messenger used by cells to regulate ion channels and gene action. Greengard studied in detail the mechanism of dopamine-mediated neurotransmission. The transmitter activates a tightly controlled series of protein phosphorylations and dephosphorylations that are influenced by cAMP. These events are supervised in turn by master regulator proteins such as DARPP-32. DARPP-32, acting like the conductor of an orchestra, directs further cascades of gene expression and receptor display.[4]

Eric Kandel was cited for his artful dissection of short- and long-term memory in a lowly animal. Much of Kandel's work has been based on studies with the marine sea slug *Aplysia*, which contains a mere twenty thousand neurons, as opposed to the over 100 billion in our own brains. Kandel found that low-intensity electrical stimuli led to short-term memory as the slug learned to withdraw its mantle (a gill-like structure) ever more vigorously. This sense of the nasty (minutes to hours) is determined by channels that move calcium into the cell, thereby increasing transmitter release. Calcium also amplifies the withdrawal reflex by phosphorylating channel proteins à la Paul Greengard. *Ohne Phosphor keine Gedanke*, indeed. In contrast, a more profound stimulus leads to long-term memory (up to several weeks) and is associated with increased levels of cAMP within the neurons. Genes are then activated via cAMP-dependent protein kinases and promoter elements called CREBs; these modulate the form and function of the synapse by turning on protein synthesis. When protein synthesis is blocked, long-term memory is ended, but the short-term memory lingers on.[5]

ANALYZE THIS

Through medical school and clinical training, Eric Kandel wanted to grow up to be Sigmund Freud; he still spends summers by the waters of Wellfleet, Massachusetts, surrounded by shrinks and scholars, rather than among the sea slug students of Woods Hole or Cold Spring Harbor. But my hunch is that were he alive today, Sigmund Freud would have wanted to be Eric Kandel. In the Vienna of 1880, young Freud was put to work on sand eels by Brücke, his mentor in physiology. Freud derived a sound respect for the nervous system of lower animals by dissecting over four hundred eels, an exercise that helped to convince him that behavior would eventually be reduced to anatomy and/or physiology. Differentiating hysterical paralysis from the organic, he explained that "Hysteria is ignorant of the distribution of nerves. . . . It behaves as if anatomy were non-existent or as if it had no knowledge of it."[6] When Eric and I were Lewis Thomas's young faculty members at NYU, Eric had just completed his first wiring diagram of the *Aplysia* nervous system, using the techniques of classic electrophysiology. In a paradigm shift, he extrapolated the responses of this simple creature to a general notion of behavior: "This shift seemed to some students of behavior a reductionist step—almost preposterous."[7] Little did he realize that he was on his way to the study of learning, memory—and of gene action. But genes contain only part of the answer to psychiatric disease. As one of the most cultivated polymaths of science, Kandel is as familiar with *Faust* as with *Phosphor*: he has taught us that our wits are shaped by time and the ambient culture: mental health and disease are polygenic.[8] Freud the neurobiologist was there before him, contending that "The action of heredity is comparable to that of a multiplier in an electric circuit, which increases the visible direction of the needle but which cannot determine its direction."[9] The master would have been proud of Kandel.

OLD GENES IN PASADENA

News of the Nobel awards for Kandel and Greengard was cause for celebration at an Ellison Medical Foundation (EMF) "Symposium on Differential Gene Expression in Development and Aging" at the California Institute of Technology.[10] Eric Kandel is one of five members of the EMF Scientific Advisory Board and Paul Greengard is a recently appointed EMF senior scholar. It's been a banner year for the study of aging, and since aging is the prime example of polygenic decay, it's been a banner year for studies of gene regulation *en group*. The talk in Pasadena was of sets of "longevity genes," of our complex response to food deprivation or oxygen stress, and the trendiest buzz this week was of genes spread on chips, chips, and more chips. Chips are microarrays of genes or their parts, stuck on solid surfaces so formed as to pick up labelled mRNAs; when the labels light up one can study the expression of thousands of genes over time.

For many of us, the big news in Pasadena was that Richard Weindruch had found a discrete packet of genes that are turned on with age and turned off again when mice are deprived of food; caloric restriction selectively blocked the age-associated induction of these genes. Since 1935 we have known that long-term caloric restriction is the only sure means of extending the life span of mammals—and of protozoa, worms, and fruit flies.[11] Laboratory rats and mice live up to 40 percent longer when fed a diet that has at least 30 percent fewer calories than usual. The animals are free of age-related disease and appear healthy in every respect except that they are generally less fertile. Even the staid *New York Times* enthused over a recent experiment in this line by Leonard Guarente: "Actuaries, put down your gloomy mortality tables and sharpen your pencils. Heirs and legatees, contain yourselves in patience. If any such drug were to work in humans the same way that this diet of 30 percent less than normal calories works in laboratory rodents, people would start enjoying a maximum life span of 170 years, most of it in perfect health."[12]

Richard Weindruch told his audience that aging in the central nervous system involves "atrophy of pyramidal neurons and synapses, decrease of dopamine receptors, accumulation of fluorescent pigments, cytoskeletal abnormalities, and reactive astrocytes and microglia [the phagocytes of the CNS]." He used microarrays of 6,347 genes and found a gene-expression profile "indicative of an inflammatory response and of oxidative stress," findings like those in the aging human brain. Caloric restriction selectively blocked the age-associated induction of these genes.[13] His audience understood that the genes involved were markers of inflammation and suggested that perhaps infections of some kind might be involved; perhaps the intestinal flora might be at fault. Perhaps with age the gut permitted uptake of bacterial products . . . perhaps . . .

AFTER MANY A SUMMER DIES THE SWAN

Craig Venter of Celera told the Pasadena crowd that it had taken his company less than six months to map the genome of three different strains of mice. Venter said that Celera was 95 percent finished by mid-October; he had again beaten out a public consortium that won't be finished until spring.[14] Since mice and humans are 85 to 90 percent identical genetically, and since Weindruch's "inflammation and stress" genes are almost identical to those of humans, Venter's work is exhilarating. The three inbred strains have permitted Venter & Co. to identify some real differences in SNPs (single nucleotide polymorphisms) that are markers for what in the human would be differences among individuals. That means we can begin to assign not only a name and a place to those inflammation and oxidative stress genes, but who's got them, in what dose, and how one would go about changing them.

But wait. I seem to have heard this song before. It's a story set in Los Angeles, where a boorish tycoon has built a large mansion (half the San Simeon of Hearst, half the Huntington Library) that houses valuable ancient manuscripts. An English scholar is sent to look into a sheaf of books and manuscripts that the mogul has acquired from an English earl and that date back over three hundred years. The tycoon has hired a certain Dr. Obispo [sic] to inject him with an extract that will prolong his life. The extract is prepared from the intestinal flora of carp—whose life expectancy far exceeds that of humans. The tale is told in *After Many a Summer Dies the Swan* (1939) by Aldous Huxley and is based on the notions of Élie Metchnikoff, who won the Nobel Prize in 1908 for discovering phagocytosis.[15] He also suggested that we age because our intestinal bacteria generate toxic metabolites of fatty alcohols that activate "neuronophages" (read: glial cells). Dr. Obispo does the carp-intestine experiment in mice: "The effect on the mice had been immediate and significant. Senescence had been halted, even reversed. They were younger at eighteen months than they had ever been." Just like the animals of Weindruch.

The locale also rings a bell. It turns out that Huxley's fictional mogul is the chief supporter of an equally fictional university, Tarzana U, about to be erected near that monumental archive of manuscripts. The founder of Tarzana, Dr. Mulge, brags to his donor that "The Athens of the twentieth century is on the point of emerging here in the Los Angeles Metropolitan area. I want Tarzana to be its Parthenon [for] Art, Philosophy, Science."[16] Tarzana seems to be the fictional stand-in for Cal Tech; Huxley in 1939 was living in Los Angeles and experimenting with hallucinogens that might spur the imagination. The seminal spirit of Cal Tech was George Ellery Hale, an astronomer and the first director of the Mount Wilson Observatory. Dean Hale foresaw the future development of his Parthenon in Pasadena: "No creative work, whether in engineering or in art, in literature or in

science, has been the work of a man devoid of the imaginative faculty."
Dr. Mulge would have been pleased that the California Institute of
Technology is within a stone's throw of the Huntington Library, where
this week is displayed a copy of *The Ellesmere Chaucer*, acquired by
Henry Huntington in 1917 from the third Earl of Ellesmere, whose
family had owned it for three hundred years.[17]

The title of Huxley's book is taken from Tennyson's poem "Tithonus," that eponymic, handsome mortal beloved by Eo, goddess of
dawn. The goddess was so enamored of the young swain that she asked
Zeus to make Tithonus immortal. He did, but neglected to prevent
Tithonus from aging. As time went on—as the genes did their stressful
work—he aged, dried, and shrank to the size of a cricket. And that's
why crickets chirp at dawn.

> The woods decay, the woods decay and fall,
> The vapours weep their burthen to the ground,
> Man comes and tills the field and lies beneath,
> And after many a summer dies the swan.[18]

Now let's get cracking on the SNPs.

October 3, 2000

RU 486 Comes to America:
Hommage à Claude Bernard

THIS IS the week that the FDA finally approved RU 486 and its prostaglandin partner, misoprostol, for oral use in first-trimester abortion. The story of RU 486 and misoprostol illustrates the adage that no distinction can be drawn between basic and applied science: there is only science that has or has not yet been applied.

RU 486 MEETS THE PRESS

When FDA commissioner Jane E. Henney announced approval of the antiprogestin RU 486 for abortion in the first seven weeks of pregnancy, the United States joined most of Europe, China, and a good part of the "undeveloped" world in receiving a "safe and effective drug," to quote Dr. Henney. However, in our God-fearing country where abortion clinics have been picketed and bombed and where doctors have been shot, opposition was quickly raised by the devout, the deranged, and the disappointed. Cardinal Egan of New York charged the agency of approving for the first time a drug designed to end, rather than to prolong, life. An agitated parish priest drove into a women's clinic wielding an axe only to be stopped at gunpoint. Speaking on *Meet the*

Press, presidential candidate Pat Buchanan called RU 486 "a human pesticide." Judie Brown of the American Life League, an anti-abortion group, cried that "We will not tolerate the FDA's decision. . . . RU-486 is a chemical assault weapon aimed at the tiniest babies."[1]

Burdensome state laws, which require disposal of fetal tissue by burial and special ventilation systems for abortion clinics, added to FDA restrictions, such as readily available ultrasound and surgical abortion, have already created a number of obstacles to the clinical use of mifepristone. If the zealots have their way, more barriers will surely be raised. Nevertheless I am optimistic that the RU 486 regimen (two 300 mg tablets of mifepristone, followed thirty-six to forty-eight hours by 200 mg of the prostaglandin analogue misoprostol) will prove to be as effective ($>$ 95 percent) and safe (again $>$ 95 percent) in the United States as it has proved worldwide where over 3 million women have had medical abortions, most without the restrictions set out in the FDA ruling.[2] It should be pointed out that mifepristone has other uses as well. It is effective as an antiglucocorticoid and may control estrogen-dependent tumors. Finally, given after a single act of unprotected intercourse, and/or once-a-month treatment immediately after ovulation, RU 486 has shown "high contraceptive efficacy."[3] The zealots are not likely to approve of mifepristone as a morning-after pill, either. As someone who witnessed the human tragedies of the rusty-coat-hanger era before *Roe v. Wade*, I would argue that the best way to handle the militants and their shenanigans is to take *all* abortion services under the wings of our large teaching hospitals. Let them picket Hopkins and the Mayo Clinic.

THE FRENCH TOUCH

The real father of the pill is Emile-Etienne Baulieu of INSERM Unit 488 (read: state-supported research unit) at the Hôpital Kremlin-Bicetre of Paris. In the 1960s and 1970s, he and I were working on rather similar problems and I've followed the work of this cultivated, elegant polymath ever since. My lab had discovered that steroid hormones such as estradiol and progesterone had direct, and opposing, effects on biomembranes.[4] Baulieu confirmed these findings, and over the next decade took them to a higher level by studying the effects of steroids on oocytes. He described specific membrane receptors for estrogens and progestins and—with collaborators at Rousell-Uclaf— went on to synthesize antagonists of hormone binding. By 1978 he ventured that "as yet there is little supportive evidence for the existence of a hormone-membrane interaction mechanism in somatic target cells, but . . . a hormone may well have more than one molecular target in a given cell."[5] He reviewed studies of cell membrane interactions with "steroids modified to prevent cell entry in order to determine whether the interactions observed in oocytes are of wider significance."

Baulieu's work became "of wider significance" when it turned out that steroids "modified to prevent cell entry"—antiprogestins such as RU 486—could counteract other effects of progesterone not only in vitro but also in life. We now know that steroid hormone receptors in the womb, as elsewhere, are assembled into a complex of several proteins, chiefly heat-shock proteins called Hsp90. Hsp90s are necessary to maintain the receptor in a conformation that can bind the hormone. When progesterone binds, Hsp90 dissociates from the steroid receptor, and the hormone-bound receptor activates genes required to nurture the egg in utero. RU 486 blocks this process.[6] Four years after the basic work was finished, Baulieu's antiprogestin was ready for the clinic.[7] And shortly thereafter, the first full account of the use of RU 486 in fertility control appeared in a prescient volume, co-edited by Baulieu and Sheldon Segal of the Population Council, an ebullient champion of reproductive rights.[8] Basic science had become applied science—at least in France. Reading the Baulieu/Segal book today makes it difficult to understand why it took so long for this safe and useful agent to arrive in America. Looking at that videotape of Pat Buchanan on *Meet the Press* makes it easy.

MISOPROSTOL: *LA SCIENCE C'EST NOUS*

The second half of our story has a gentler tone. In the dark days after the market crashed in 1929, a Brooklyn woman consulted her gynecologist, Charles Lieb, because she experienced severe pain each time she had intercourse with her husband. Reasoning that there might be something in the husband's seminal fluid that caused uterine contractions, Lieb teamed up with a another young gynecologist, Richard Kurzrock. The two went to the laboratory of Sarah Ratner, their friend and a very junior pharmacologist at Columbia, to see if human semen had in it anything that might contract human uterine muscle. They attached strips of muscle obtained at operation to a strain-gage attached to a stylus which made scratches on a charcoal-smoked drum. Sure enough, semen did contain a bioactive compound that made the muscle contract. Those scratches on the drum were the birth pangs of prostaglandins.[9] When I was a medical student, Sarah Ratner, by then a professor of pharmacology with Severo Ochoa at NYU, was still coating her own recording drums in charcoal and was still finding new bioactive compounds.

The Kurzrock and Lieb paper was forgotten for decades until the prolific Swedish physiologist Ulf S. von Euler put his student, Sune Bergström, on the trail of bioactive substances in sheep organs; soon enough a substance was found that contracted uterine muscle: it was called "prostaglandin" and by the time Bergström put *his* student, Bengt Samuelsson, on the trail, it was clear that the material came not from the prostate gland itself, but from arachidonic acid precursors in the seminal vesicles.[10,11] Based on Samuelsson's and Bergström's chemical analyses, the prostaglandins were isolated and defined, their roles in the widest range of physiologic and pathologic conditions were defined, and in 1982 Nobel Prizes were gained by Bergström, Samuelsson, and John Vane—who showed that aspirinlike drugs inhibited the biosynthesis of prostaglandins.[12] Misoprostol (a prostaglandin E surrogate) was developed by Searle in order to overcome aspirin's inhibition of prostaglandin biosynthesis in the stomach, an effect that neatly explains why aspirinlike drugs cause gastric irritation. It was also clear that E-series prostaglandins, acting via a family of discrete surface receptors (EPRs), provoked contraction of human uterine muscle, and did so by raising cyclic AMP. In the RU 486 drill, the uterus is deprived by mifepristone of progesterone for thirty-six to forty-eight hours, then prostaglandin comes along to empty the womb.

This week, women in America will be able to share with their sisters the world over the benefits of basic science translated into human choice. It's a good time to remember the community of basic and applied scientists that made it all happen: Baulieu and Segal, Kurzrock and Lieb, Bergström and Samuelsson, Ratner and Vane. "L'art c'est moi; la science c'est nous" claimed Claude Bernard in 1865.[13] Right on.

September 18, 2000
Leptospirosis, Tularemia, and
Prince Kropotkin

THIS WEEK Darwin and Marx reared their nineteenth century heads. The fittest of yuppies came down with Weil's disease (leptospirosis) after surviving a made-for-television race in Borneo, while workmen caught tularemia from cute little rabbits in Martha's Vineyard.

ECO-NARCISSISM

On September 14, the *New York Times* reported breathlessly that 30 of 155 Americans who participated in the seventh Eco-Challenge expedition race in Malaysian Borneo had become ill from leptospirosis "and many have been hospitalized."[1] The CDC took action after being notified of several particularly severe cases.[2] The twelve-day race, through caverns, canyons, swollen rivers, and guano heaps, was a mecca for corporate eco-narcissists. Most of the seventy-six four-person teams were composed of thirty-something fitness instructors, bond salesmen, policy analysts, and others of the triathlonic persuasion. The teams bore the logos of the global economy, from Salomon/Eco-Internet to Advil Canada; from Playboy Extreme to Ernst and Young. We ain't talking Médecins sans Frontières, here. Indeed, a four-hour documentary of the race is scheduled for the USA network next spring, and their web-site mugshots look like the contestant pool for NBC's *Survivor* show.[3] It is no coincidence that Mark Burnett, who founded Eco-Challenge as an extreme sports competition "promoting a positive environmental mes-sage," is in fact the executive producer of *Survivor*. Following the CDC alert, Burnett posted an ingenuous notice to his fellow social Darwin-ists: "Unfortunately, the jungle environment took its toll on this year's Eco-Challenge; there are some competitors who have become ill after returning from the Borneo race. To those of you who are sick, I wish you all a very speedy recovery. To those who are home, I recommend a routine checkup with your doctor."

Leptospirosis is no laughing matter.[4] Mortality rates can range as high as 25 percent from liver and kidney failure; pulmonary hemor-rhage—due to disseminated intravascular coagulation—is a dread complication. Weil's disease is caused by a spirochete, *Leptospira inter-rogans*, that lives in the urinary tract of rodents, dogs, bats, and other creatures of the wild; their infected urine spreads the microbe. Lep-tospirosis is endemic in Southeast Asia, Oceania, and India. Epidemics have followed flooding from Latin America to Bangladesh. When I was chief medical resident at Bellevue many years ago, our Weil's dis-ease patients were Hispanic kids who jumped from dirty, rat-infested docks into the East River to cool off from August in the city. Lately, the disease has spread to posher venues: triathletes in Illinois and Wiscon-sin, corporate rafters in Germany, and—in Malaysia itself—to "The patient who had a picnic at a waterfall and presented with haemoptysis and renal failure."[5, 6, 7]

IT'S A RAINFOREST OUT THERE

Several of the Eco-Challenge victims were very sick, indeed, with fevers of 104°F and meningismus along with the usual liver and kidney involvement. The disease was unexpected. Burnett's prospectus told the competitors that they would: "navigate through ancient caves and paddle along winding rivers in indigenous sampan canoes. . . . Teams will trek and mountain bike along dense rainforest trails." Instead, the course took them across rivers flooded by unending rain, leech-infested ponds, guano in the caves, and mudslides under the "rainforest canopy." One notes that in the new language of eco-speak, it's "rain-forest" to the laity, it's only a jungle out there when you get sick—or the mike is off. Reporting on the site in March of this year, the Eco-Challenge advance party described their helicopter skimming "the seemingly impenetrable ceiling of the jungle [*sic*] as we approached the site of our first adventure. Brown and rapid, the rivers were swollen from a month of relentless rains across the island." Not only in March, but also in August during the race: "The Taliwas River was more like a creek early yesterday, and crossing it was simply a matter of wading through ankle-deep water. But then the rain fell, the water rose and the river became impassable. A mini-waterfall appeared where there had been none before."[8] A perfect place for that picnic by a waterfall—and for leptospirosis.

It wasn't that bad for all the participants. One of the winning four-some, Isaac Wilson of Salomon/Eco-Internet, came across a group of elephants and confessed that "We literally stopped biking to watch them"—a remarkable feat for this fan of the wild whose team furi-ously outpedalled its fellow entrepreneurs. Another eco-epiphany was that of his teammate, Robyn Benincasa, who reported on-line: "About two in the morning last night, we're paddling and I turn around in the *perahu* [native sail canoe] and there are three naked guys behind me, just airing it all out. So I thought, well, I just have to join them." She called it their "naked paddling episode" on the way to victory. Out there in Salomon/Eco-Internet land it's Nature raw in tooth and claw where only the fit survive.

RABBIT FEVER

Leptospirosis may be a recreational illness in Borneo, but tularemia is an occupational disease on Martha's Vineyard. Sounds like Marx, not Darwin. This week a CDC team arrived on the island to find out exactly why David Kurth, a forty-three-year-old house painter and yardworker, died of pulmonary tularemia and why nine other "landscapers" (read: "garden help") caught the often lethal microbe.[9] The proximal cause of this airborne disease is the inhalation of rabbit dung bearing the small gram-negative rod *Francisella tularensis*; other forms come from tick bites or uncooked rabbit entrails. The underlying causes are a mixture of history, biology, and sentiment; the same mixture also brought Lyme disease to Cape Cod and the Islands.

The yard-help business on the Vineyard involves a lot of brush- and thicket-clearing, according to locals: "After all, here on the Vineyard, city dwellers tend to let their yards run to meadow and don't think of rabbits as vermin."[10] There was no tularemia in Massachusetts until the 1930s after nearly thirty thousand rabbits were imported for sports hunting, and it is the descendants of those rabbits—now no longer permitted to be hunted or trapped—that frolic around the meadows and brush. The disease is maintained "endozootically" by the dogtick. That parallels the entry and spread of Lyme disease in New England. By the late nineteenth century, deer had been essentially eliminated from the area by Indian and Yankee deerslayers; brush and forest had been cleared for farmland. But the wealthy Forbes family, who owned Naushon Island, a private enclave within deer-swimming distance of the Cape, imported deer for their sport. (Teddy Roosevelt was a frequent visitor.) The progeny of the Forbes' Bambis—now no longer permitted to be hunted or trapped—became the winter haven of *Ixodes dammini*. When deer are eliminated from an area, Lyme disease effectively disappears.[11] No such experiment has been done with bunnies.

ECO-SENTIMENTALISM

But the mere presence of cute little rabbits, with their tularemia and dog ticks, or those fetching deer, with their cargo of *Ixodes*, is not sufficient to spread the disease. Andrew Spielman of Harvard has pointed out that deer became abundant on the Cape when summer homes replaced farms as the chief housing stock: "Residential development seems to favor small tree-enclosed meadows interspersed with strips of woodland, a 'patchiness' much prized by deer, mice, and humans," not to speak of rabbits.[12] As a result, increasingly large numbers of people live where the risk of Lyme disease or tularemia is high. On the Vineyard, employees, not employers, contract tularemia. Public health authorities have warned landscapers—but not the owners of summer homes who on the Vineyard are not expected to do their own yard-work—to wear masks and gloves in the field. Reason dictates that if we want to eliminate tularemia we need to kill those little varmints, or give up our vacation homes, our meadows, and our brush. Fat chance. We're more likely to settle for the eco-sentimental solution, and do neither. More workers on Martha's Vineyard are likely to come down with tularemia and some will die. The rabbits are taking care of each other; perhaps the Vineyarders will also band together for mutual assistance.

Considering the problem of mutual aid, my thoughts turned to the father of eco-biology, the anarchist Prince Pyotr Kropotkin (1842–1921). The prince was the first to draw attention to animal altruism based on his field observations in Siberia. For him, the rabbit was the very model of the social animal, whose persistence—and capacity for mass production—was an argument against the dog-eat-dog world of ordinary (social) Darwinism. The rabbit, he told his friend Ford Madox Ford, stood out against the fiercer aspects of selection. Defenseless and adapted for nothing in particular, it had outlived "the pterodactyl, the Hyrcanian tiger and the lion of Numidia."[13] Anticipating E. O. Wilson, he used the rabbit as an example of "Mutual Aid," by which survival of the species is not necessarily the result of a Hobbesian domination of the weak by the stronger.[14] In his *Memoirs of a Revolutionist* (1899), he complained that the Darwinian struggle for existence had been misinterpreted in the biological, as well as the social, realm. He cried that "there is no infamy in civilized society, or in the relations of the whites towards the so-called lower races" that has not been excused by the Darwinian "struggle for existence."[15]

The prince was no eco-sentimentalist: Altruism had survival value; it had nothing to do with "love" in its sentimental sense:

> It is not love, and not even sympathy (understood in its proper sense) which induces a herd of ruminants or of horses to form a ring in order to resist an attack of wolves; not love which induces wolves to form a pack for hunting; not love which induces kittens or lambs to play, or a dozen of species of young birds to spend their days together in the autumn; and it is neither love nor personal sympathy which induces many thousand fallow-deer scattered over a territory as large as France to form into a score of separate herds, all marching towards a given spot, in order to cross there a river. It is a feeling infinitely wider than love or personal sympathy—an instinct that has been slowly developed among animals and men in the course of an extremely long evolution, and which has taught animals and men alike the force they can borrow from the practice of mutual aid and support, and the joys they can find in social life.[16]

Rabbits have survived, and deer have persisted by mutual aid, and the diseases they carry have also persisted. In Kropotkin's sense, it's not for love that the Vineyarders will eventually have to kill the varmints.

September 5, 2000
Pope Says No to Cloning

THIS WEEK all roads led to Rome, where the International Congress of the Transplantation Society became the center of a storm over cloning and stem cell research.

VATICAN V. DOLLY

Pope John Paul took the unusual step this week of leaving his cool summer palace, the Castel Gandolfo north of Rome, to address a scientific meeting in the steaming city. The eighty-year-old pontiff, showing the effects of his losing battle with Parkinson's disease, lifted his frail, incantory voice to thank the audience of transplant experts for having developed a technique that permitted donors to engage in an "act of love." "Transplants are a great step forward in science's service of man, and not a few people today owe their lives to an organ transplant," said the pope. That tribute won repeated applause from his audience of about four thousand specialists, as did the homage he paid to such newer techniques as bone marrow transplantation and adult stem cell research.[1] His audience was somewhat less than enthusiastic when the pontiff reiterated a recent Vatican directive forbidding research on embryonic stem cells, human fetuses or what he called the "frutto della generazione umana" (the fruits of human generation).[2]

The pope was quoting a text prepared by his Pontificia Accademia Per La Vita (Papal Academy of Life) which, on the occasion of the pontiff's address, had issued its *Dichiarazione Sulla Produzione E Sull' Uso Scientifico E Terapeutico Delle Cellule Staminali Embrionali Umane* (Directive on the production and the scientific and therapeutic uses of human embryonic stem cells). The document, soundly argued and based on good modern science, takes a no-nonsense view of what human life is, and when it begins.[3] The directive argues that all the "fruits of human generation" should be guaranteed "the unconditional moral respect deserved by the human condition, both physical and spiritual" (rispetto incondizionato che è moralmente dovuto all'essere umano nella sua totalità e unità corporale e spirituale)." In accord with that position, the pontiff insisted that methods that fail to respect the dignity and value of the person must always be avoided.[4]

The cloning empire struck back. Speaking to Italian reporters immediately after the Pope's remarks, Dr. Ian Wilmut, impishly headlined as "Il papa di Dolly" (the father/pope of dolly) by the *Corriere della Sera* (Milan), argued persuasively that "un embrione non è ancora una persona" (an embryo is not yet a person). He pointed out that an embryo is only a potential human being, since it lacks a nervous system; therefore no ethical barriers should be raised against those who wish to use embryo cells for research or treatment.[5] Indeed, both in the United States and in Great Britain, research on embryonic stem cells has been given a grudging go-ahead by the responsible authorities.[6]

STEM CELLS, CLONING, AND THERAPY

Let's define what the two papas were debating. In **embryonic stem cell cloning,** a fertilized egg is permitted to reach a stage somewhere between blastula and gastrula (that is, about a hundred to a thousand cells). From this round homuncular cluster one can extract an inner cell mass of pluripotential cells. Given appropriate culture conditions in the dish or in a recipient, such cells can form any cell in the body except, perhaps the organs of special sense and, certainly, the placenta. Ironically, given the pope's own medical condition, one of the most promising applications has been the successful treatment of Parkinson's disease by stem cell therapy.[7]

Somatic cell cloning is quite different. A nucleus from an adult somatic cell (diploid) is taken from any tissue of a donor and plonked into an unfertilized ovum of the same species whose own nucleus has been removed. If the egg cytoplasm acts properly on the genes of that strange nucleus, if the growth factors are favorable, and if the moon is right, a whole new adult can be produced. But only after the creature has been reinserted into a uterus of the same species and permitted to come to term. That's how Dolly was made and that's cloning.[8] The Brits and the Clinton administration have forbidden research that would introduce such a cloned homunculus into a uterus.

In **therapeutic somatic cell cloning,** the egg with its foreign nucleus is kept in vitro. The resultant assemblies are cultured in another defined brew of nutrients until—with time, luck, and a pinch of tissue-specific hormones—they can be persuaded to become a pound of the proper flesh: islet cells for diabetes, liver cells for cirrhosis, and, again, brain cells for Parkinson's disease. But the new molecular biology will permit us to doctor the genes of those cells at will. To quote the papa of Dolly: "Precise genetic modification will be achieved by site specific recombination in the donor cells before nuclear transfer. In all mammals it will become possible to define the role of any gene product and to analyze the mechanisms that regulate gene expression."[9]

SAYING NO TO CLONING

These sci-fi prospects have raised fears in the minds of others than the gentle pontiff. In his jeremiad against the temper of our times, my teacher and exemplar, Jacques Barzun, faults biomedicine for inducing "psychic disarray by the manipulation of genes. Cloning was only the apex of disturbing procedures."[10] The church in America is also distinctly disturbed. "For the first time in history, our federal government will promote research in which developing human beings are destroyed," said Richard Doerflinger, associate director of the Pro Life Secretariat at the National Conference of Catholic Bishops.[11] Without a bow to telomeres or telomerase, C. Ben Mitchell, a bioethics consultant for the Southern Baptist Convention, the nation's largest Protestant Church, intoned that "Lurking behind the scenes is a not-so-subtle quest for immortality.... A cottage industry is developing to ward off not just age related diseases, but aging itself."[12] The clearest rejection came from Senator Sam Brownback, Republican of Kansas, an area not known for evolutionary zeal. Senator Sam told reporters that human embryonic stem cell research was "illegal, immoral and unnecessary."[13]

THE POPE IS RIGHT

The debate between the two papas hinges on the question "When does life begin?" On the West Side of Manhattan, where I was raised, it was generally agreed that life begins when the fetus graduates from medical school. Be that as it may, if human life begins when a haploid sperm meets a haploid egg and a diploid blob develops in dish or womb, if one believes that all life deserves what the papal academy has called the *rispetto incondizionato*, then it follows that research that disrupts any diploid assembly will violate that respect. But wait. What about organ transplantation? Isn't that just another name for engrafting an organized blob of diploid cells? What about marrow donation? Marrow cells are simply a collection of early diploid cells that—given the right moon—can reassemble to become a whole human. Give a pint of blood, and you enter into what Richard Titmuss has called the Gift Relationship.[14] A blood or organ donor is, literally, a philanthropist who passes on the "fruits of human generation." The pope is therefore right to praise organ transplantation and marrow transplants as "acts of love." Clearly, the donor of an egg that harbors a foreign nucleus, or one that is fertilized in vitro to yield a pluripotent cell line, has also engaged in an "act of love." Since we now know that adult stem cells from one organ can turn into entirely new tissues in the dish—brain into blood, blood into brain (see next chapter)—why should one batch of diploid cells be defined as "life" while another batch is not? The papa of Dolly really has it right: none of those batches of diploid cells has a nervous system. That is formed only after the embryo develops in a uterus where two lives remain intertwined until parturition. At term, life begins in a painful act of love.

August 22, 2000

Genes, Brains, and Jacques Loeb

THE NEWS this week sounds like science fiction: Genes aren't made of DNA alone, and one can turn bone marrow into brain by marinating it in chemicals.

FASTER THAN THE SPEED OF LIGHT

Last week, Shiv Grewal of Cold Spring Harbor stood on a platform at Woods Hole and, on the basis of his work with fission yeast, announced dramatically that "The 'gene' in this instance thus comprises DNA plus the associated Swi6-containing protein complex." Put simply, he had found that in a predictable, heritable way, a specific protein in eukaryotic cells functions together with DNA as the unit of genetic information. This particular protein, called Swi6, determines which stretches of DNA itself can or cannot be copied. The DNA-protein complex, not DNA alone, dictated to succeeding generations of cells exactly how to twist and turn their chromosomes by tagging such critical areas as centromeres, telomeres, and mating sites. Normal cells lose telomeres (the ends of chromosomes) as they age; cancer cells maintain their telomeres forever. Proteins like Swi6, which control sites at which chromosomal DNA are duplicated, may be the key to the chromosomal aberrations associated with both aging and cancer.

That astounding finding, published this week in the *Journal of Cell Physiology*, puts to rest the central dogma—or mantra, if you will—of twentieth century biology that "DNA makes RNA makes protein."[1] The news was given at a colloquium on aging sponsored by the Ellison Medical Foundation. Yes, that's *the* Larry Ellison of Oracle and ocean-racing fame, and candor requires the confession I'm on the advisory board of the foundation.[2] Since both aging and cancer are associated with chromosomal aberrations—the best example being that of the Philadelphia chromosome in chronic myelogenous leukemia—everyone in that distinguished and well-funded audience knew exactly what the implications were: genes are more than the sum of their bases. Dogma was down another notch.

Central dogma had already suffered a blow in the 1970s when retroviruses—of which HIV is but one example—were discovered by Howard Temin and David Baltimore; the script was promptly rewritten to read RNA makes DNA makes RNA makes protein. It got more complicated when in the 1980s Stanley Prusiner discovered that misfolded proteins, prions, could propagate to infect cells and cause scrapie, mad cow disease, and Kuru; we were forced to admit that misfolded protein makes misfolded protein makes ... Now Grewal has recast the mantra to write that DNA-protein makes DNA makes DNA-protein makes—whatever. Let's hear it for fission yeast.

Speaking of fission, and its noble base, $e=mc^2$, Grewal's announcement was made exactly one month after physicists at the NEC labs in Princeton showed that it was possible to exceed the speed of light.[3] Wang and company had broken the cosmic speed limit, causing a twin laser-driven light pulse to travel in a cesium chamber so fast that the peak of the pulse left the chamber before it had finished entering it. Their finding of "superluminal light propagation" challenged the final constant of twentieth century physics, the c of $e=mc^2$. All this in July and August of 2000. And while I'm ill prepared to judge whether Grewal's intricate yeast genetics have indeed toppled Crick and Watson, or whether the laser optics of Wang et al. have overturned Einstein, I'm thrilled by the new pace of discovery, which seems to be moving even faster than the speed of light. An old limerick of Cambridge origin comes to mind:

There was a young lady named Bright

Whose speed was far faster than light

She set out one day,

In a relative way.

And returned home the previous night.

DOWN TO THE MARROW

By and large, articles published in the *Journal of Neuroscience Research* tend not to attract public attention: we ain't talking *NEJM*, *Nature*, or even *JAMA*. But this week, news taken from a *JNR* publication spread across the country from New York to Los Angeles: SCIENTISTS FORM BRAIN CELLS FROM BONE MARROW.[4] And that's exactly what a team led by Ira Black, a neurologist from Robert Wood Johnson Medical School in New Jersey, and Darwin Prockop, of MCP Hahnemann in Philadelphia, had done.[5] Prockop's laboratory adventures have taken him from gout to the gene therapy of osteogenesis imperfecta, the Toulouse-Lautrec disease.[6,7] Recently he has been culturing bone marrow stromal cells and getting this stem cell-like population to grow and propagate when injected into experimental animals. Since it was recently discovered that stem cells from rat brain—yes, there are stem cells there—could differentiate into blood cell precursors in vivo, Prockop and Black looked for a way to reverse this process: they have turned human blood cells into brain cells, and they found a way to do this reproducibly and by means of very simple chemicals.[8]

Black told reporters that the process is "1 percent insight and 99 percent luck and trial and error.... These and other stem cells are far more flexible than we had thought a few years ago.[9] It means that we have to rethink what we learned in medical school: that the fate of a cell is fixed." Black knew that Prockop's bone marrow stromal cells can be greatly expanded in the dish and induced to differentiate into several mesenchymal cell types: cartilage, fibroblasts, etc. However, no one had gotten such cells to differentiate into nonmesenchymal cells such as brain, heart, or lung. They achieved their goal by soaking bone marrow stromal cells in the usual witches' brew of insulin, cortisol, and growth factors, but the critical ingredient turns out to have been the antioxidant/solvent mixture of beta-mercaptethanol and DMSO. Lo and behold, for over twenty passages, over 80 percent of them were found to "exhibit a neuronal phenotype" expressing neuron-specific cell markers, and looking very much like true-blue neurons in the dish: "The refractile cell bodies extended long processes terminating in typical growth cones and filopodia."[10] Some clonal cell lines, established from single cells, proliferated and showed that they had inherited all the genetic signals required to remain neurons forever. Indeed, Black told the press that when he injected these blood-into-brain cells back into animals, the cells can survive within the spinal cord for well over a month.[11] "That tells us that they are user-friendly in the live animal." The *JNR* article modestly concluded that "adult marrow stromal cells . . . may constitute an abundant and accessible cellular reservoir for the treatment of a variety of neurologic diseases."

This discovery has two important implications. First, of course, is the finding that we each have within us an "abundant and accessible" source of stem cells. And since research on human fetal stem cells— the most obvious source of pluripotent cells for Parkinson's disease, Alzheimer's disease, or even liver regeneration—is under attack by fans of various ethical and religious causes, Black and Prockop rightly claim that: "autologous transplantation overcomes the ethical and immunologic concerns associated with the use of fetal tissue."[12] Secondly, if a simple, off-the-shelf chemical (beta-mercaptethanol) is the secret key to differentiation in a dish, it seems likely that by simply browsing in the chem room we'll find stuff to add to the culture brew that will permit stem cells to grow into kidneys, liver, or lungs. Ah, the wonder of simple chemicals.

But wait, I seem to have heard this song before. Almost exactly a century ago, and almost exactly in the same spot where Shiv Grewal made his announcement on fission yeast, Jacques Loeb, the founder of general physiology and the model for saintly Dr. Gottlieb in Sinclair Lewis's "Arrowsmith," sounded a similar note. Loeb reported that he had fertilized a sea urchin egg in the absence of sperm by varying the ionic composition of medium: cells could be diverted to new path dish by adding simple chemicals to a dish. The headline in the *Boston Herald* could have applied to the experiments of Black and Prockop—or to Grewal's: "CREATION OF LIFE. STARTLING DISCOVERY OF PROF. LOEB. LOWER ANIMALS PRODUCED BY CHEMICAL MEANS. Process May Apply to Human Species. Immaculate Conception Explained. Wonderful Experiments Conducted at Woods Hole."[13] And sure enough, fans of the natural order accused him of toppling "the whole structure of our ideas of life."[14]

These lives of the cell—from the dish to the marrow—remind me that without Loeb and simple chemicals we would not have among us the children produced by *in vitro* fertilization. And without further research on human embryos, who knows what—or whom—we will be missing a century from now.

August 8, 2000
Nicotine and Marijuana;
Auden and Ginsberg

KNOCKOUT MICE AT THE MBL

FRIDAY NIGHT lectures at the Marine Biological Laboratory at Woods Hole are preening displays of modern science. At their best, they combine the rigor of New York's Harvey Society with the flair of a Paris fashion show. Last Friday at the MBL, a jovial Jean-Pierre Changeux of the Institut Pasteur and the Collège de France told the exciting story of how nicotine affects the brain.

The news was good and bad: on the one hand, nicotine improves memory and spatial orientation, promotes synaptic connections, and slows aging. On the other hand, its addictive properties are written inexorably in the genes of mice and men. These facts were established by means of elegant physiologic and behavioral studies of mice rendered genetically deficient in one or another of the subunits of the nicotinic acid receptor for acetyl choline in the brain.[1] These knockout mice revealed what Changeux described as "unexpected pharmacological properties." The receptor, a ligand-dependent ion channel, is composed of five subunits. Two of these subunits are required for such nicotine-elicited responses as indifference to pain, while an entirely different one is responsible for nicotine addiction; yet a third subunit protects mice against chemical insults. Alzheimer's patients have trouble moving acetylcholine to these receptors, and anything like nicotine, which can make up for this deficit by hooking up to these receptors, might make sense as treatment. On the basis of differences in the responses of young and old knockout mice to nicotine, Changeux modestly concluded that "these animals might serve as a useful animal model for some aspects of aging, particularly those associated with dysfunction of the neurotransmitter system for acetyl choline, as seen in Alzheimer's disease." Conclusions from knockout mice may not extend to humans, but there is ample evidence that nicotine improves memory, alertness, and our performance of intellectual tasks.[2] At least one clinician in Changeux's audience thought of telling his Alzheimer's patients to resume smoking, but worried that they'd forget where they left that butt.

But before you go out and get grannie her next cigar, take a look at cautionary data that were published this week on both sides of the Atlantic. It presents the other side of the nicotine mirror. Oxford's epidemiology unit, one of the world's best, reported the results of a prospective study of smoking in England. Their message is loud and clear: tobacco kills and if we stop smoking we preserve life.[3] Some of the news from Oxford was expected: in middle-aged men (thirty-five to fifty-four) the prevalence of smoking halved between 1950 and 1990 and the death rate from lung cancer fell even more rapidly. Sadly, women and older men who were still current smokers in 1990 were more likely than those in 1950 to have been persistent cigarette smokers throughout adult life and so had higher lung cancer rates than current smokers in 1950. The good news from Oxford was that it pays to stop smoking even after decades of addiction; indeed, current efforts directed chiefly against smoking by the young may need redirection. People who stopped smoking even well into middle age (sixty, *sic*) avoided most of their subsequent risk of lung cancer. It was good to hear that those who stopped smoking at sixty cut their risk of tobacco-related death by age seventy-five to more than 90 percent. Peto and company pointed out that campaigns directed against smoking among the young would not be expected to decrease the overall bills of mortality until the middle or end of the twenty-first century. In contrast, mortality in the near future could be substantially reduced if current smokers kicked the cigarette habit now, once and for all. Would switching to cigars help? No. A new consensus review by the American Cancer Society warns us again that smoking cigars instead of cigarettes does not reduce the risk of nicotine addiction (remember that addiction-determining subunit of Changeux's receptor?).[4] Sounding a warning to all those cigar aficionados, the ACS assures us that "as the number of cigars smoked and the amount of smoke inhaled increases, the risk of death related to cigar smoking approaches that of cigarette smoking."

Remember those photographs of Ulysses S. Grant, Sigmund Freud, or H. L. Mencken with their cigars, W. H. Auden, Lillian Hellman, or Albert Camus with their cigarettes? Cancer and emphysema stare from their faces. The upside of those nicotine/receptor subunit interactions may have been mental acuity, the downside was sickness and the grave.

JOINT EFFORT

Marc Feldmann of the Kennedy Institute for Rheumatology in London is an unlikely candidate for *High Times*'s hero of the month. He is a scholarly immunologist who has won many glittering prizes; he was among the first to pinpoint tumor necrosis factor (TNF) as a culprit in rheumatoid inflammation.[5] What's a nice guy like that speaking up for pot? "A plant is a complicated mixture of parts and some can be used for good," Feldmann told the *New Scientist* on August 5.[6] In what might be called a joint effort—with scientists at the Hebrew University—Feldmann found that one of the major components of the five-pointed weed, cannabidiol, is a very effective anti-arthritic agent in experimental arthritis.[7] The model Feldmann et al. used is that of an immune complex-induced disease in which animals make antibodies against their own or related Type II collagen to start the parade of TNF-associated inflammation. The antigen-antibody complexes, like those present in human disease, activate complement and, in turn, cause lymphocytes to proliferate. Phagocytes are then turned on to release their powerful oxidants. Feldmann et al. found that cannabidiol stopped the disease dead in its tracks even when launched, not a common finding in this model. Finally he may have gone overboard—but not escaped attention from the press—when he wrote, "The present study shows that cannabidiol, a natural constituent of marijuana, is effective as an anti-arthritic therapeutic in established collagen-induced arthritis. Its efficacy *when given orally* renders it an attractive candidate for the treatment of rheumatoid arthritis (my italics)." Wow.

But wait a minute. Isn't this the mirror image of another story? Didn't we come across something like this before? Sure enough. In July of 1998, an NIH group including Julius Axelrod (Prix Nobel, 1970, for synaptic transmission) found that cannabidiol acted as a powerful antioxidant, and protected brain cells against several sorts of chemical injury.[8] The product from pot was even better than vitamins C or E as an antioxidant and the NIH press release bragged that cannabidiol holds promise for preventing brain damage in "stroke, Alzheimer's disease, Parkinson's disease, and perhaps, heart attacks."[9] Every medical news service and columnist in the country picked up the story; pot-heads and eggheads agreed that stroke and heart attacks were about to succumb to cannabinoids—all based on tests of antioxidant activity in a dish. Those of us who remember the rambles of the late Allen Ginsberg in his later years might worry more about other effects of the cannabinoids. Nothing new on this since then.

So here we are again, two years later, with news that cannabidiol is effective as an anti-arthritic therapeutic. Aren't we lucky that Alzheimer's disease and rheumatoid arthritis are about to vanish in a puff of smoke?

July 27, 2000

The Mediterranean Diet

THE MEDICAL NEWS this week seemed like a page from A. J Liebling, the Hemingway of cuisine, who put fine food on the literary map of America at midcentury.[1] Since then, we've learned that the Mediterranean diet he championed is really good for us, but we didn't exactly know why. New studies on olive oil and red wine justify Liebling's gentle advice that since those who love the pleasures of the table have but two opportunities a day for fieldwork, they are not to be wasted worrying about cholesterol.[2]

OLIVE OIL IN FLORENCE

Some weeks ago, several comrades of the prostaglandin campaigns were sitting around a table in an olive grove on a hill above Florence at sunset. We were embarked on our second round of fieldwork for the day. Piero Mannaioni, the professor of pharmacology at the University of Florence, was telling us that the greatest miracle of preventive medicine ever developed was standing right before us, ready to go into the salad, over the pasta, a bath for the fish. "Take that bottle of olive oil," he said, "It's really fine, first-pressing, virgin oil. You can leave it on a table for weeks or for months, it won't go bad. It's got so many antioxidants that it won't go rancid like butter. Besides, oleic acid itself is good for you." We tucked in, cholesterol be damned.

The minute I got back home, I checked out "olive oil" on *PubMed*, the internet source of all medical wisdom. Mannaioni was right. We've learned from well-controlled epidemiologic studies that a Mediterranean diet, rich in olive oil, protects our hearts from the slings and arrows of outrageous thrombi; it also seems to protect us from cancer of the gullet and stomach.[3] But no one was certain what the major beneficial ingredient was. Two studies this week suggest that the hero is oleic acid, a C18 mono-unsaturated fatty acid, which constitutes 70 percent of olive oil.[4] Oleic acid is the key player in the Mediterranean diet, and new studies have shown exactly how it affects discrete biochemical pathways in the cell. It's amazingly good at protecting DNA from mutilation and dampening the kinase cascades that lead to inflammation and cancer.

In this week's issue of the *International Journal of Cancer*, a group of Professor Mannaioni's colleagues in Florence report on the relationship between diet and DNA damage in Italy.[5] They studied DNA adducts in peripheral leukocytes of healthy adults; these adducts are reliable indicators of exposure to toxic agents and subsequent risk of cancer risk. Prospectively, they questioned 47,749 men and women, thirty-five to sixty-four years, in five centers, as to their individual dietary and lifestyle habits. Of these, a hundred volunteers were randomly selected from each of the three main geographical study areas (northern, central, and southern Italy). And, sure enough, the Florentines reported, the more olive oil consumed, the more antioxidants and oleic acid consumed, the fewer DNA adducts were found. These data confirmed earlier clinical studies on esophageal cancer in Italy and on prostate cancer in Greece.[6,7] In each study, oxygen-derived free radicals generated by carcinogens or by the wear and tear of aging were held responsible for the damage to DNA, and in each study olive oil was protective.

Olive oil not only protects DNA from the slings and arrows of outrageous radicals, but it may keep our minds alert. In the next chapter,

I'll describe mice that had been bred to develop an Alzheimer's-like disease and were returned to normal learning by a protein vaccine. This week we have news of normal mice who became smarter thanks to a protein activated by—no fooling—intracranial olive oil.[8] Routtenberg has been studying a protein called GAP 43, which is highly expressed in the brain of young animals and which mediates the connections that the developing brain makes in infancy (neuronal plasticity). Years ago, Routtenberg had found that oleic acid, or olive oil itself, activated GAP 43.[9] And now, when he and his Swiss collaborators created a strain of mice that overexpressed the gene for GAP 43, the animals did much better than controls in mazes used to test rodent intelligence. And while Routtenberg admitted that genetic engineering with GAP 43 wasn't quite ready for prime time in humans, he allowed that its activator—olive oil—might serve in the meantime. I might note that the medication is available over the counter.

TAKE TWO GOBLETS AND CALL ME IN THE MORNING

The other part of the Mediterranean story is even more fun. For a few years now, it's been clear that grape skins, as concentrated in red wine, contain a unique chemical, trans-resveratrol (or trihydroxystilbene), which has important anti-tumor and anti-inflammatory effects. The grapes use resveratrol to protect them against fungi, but it has more interesting effects in humans. This spring, we learned that resveratrol—at concentrations present in the blood after two glasses of red wine—blocked TNF-induced activation of NF-kappa B in human white blood cells: NF-kappa B is the key transcription factor that turns on the genetic machinery common to both tumor induction and inflammation.[10] Moreover, resveratrol can block neutrophil activation, clumping, and free radical generation.

In this month's issue of *Cancer Research*, scientists at the University of North Carolina pinpointed exactly how resveratrol inhibits activation of NF-kappa B.[11] It turns out that the site of action of resveratrol in the NF-kappa B cascade is the very same site that is inhibited by anti-inflammatory levels of salicylates or aspirin. By inhibiting breakdown of NF-kappa B's inhibitor, resveratrol—like sodium salicylate—puts the brake on the whole machine. Simultaneously with the North Carolina report, J. J. Moreno from the University of Barcelona found that resveratrol inhibited prostaglandin synthesis mediated by either COX-1 or COX-2; the resveratrol in two glasses of red wine were as powerful as celecoxibid or ibuprofen in blocking prostaglandin release.[12] It may not be an accident that resveratrol from the grapevine acts at the same intracellular target as salicylate from the willow: NF-kappa B. But it *is* remarkable that the same molecule has been shown to be the site of action of prednisone and other anti-inflammatory steroids.[13] Who would have thought that two glasses of red wine would be the ideal anti-inflammatory agent, combining the better properties of aspirin and cortisone.

We are such stuff as wine is made on.

July 20, 2000

Alzheimer's Disease and
City Hospitals

THE NEWS this week was of mice and men with Alzheimer's disease and it came from the White House. A different source, *US News and World Report*, released its annual ranking of American hospitals. Both stories confirmed Henry James's contention in *The Golden Bowl* that science is the absence of prejudice in the presence of money.

ONE FOR THE GIPPER

On July 16, President Clinton announced that federal researchers will get an additional $50 million over the next five years for research into the prevention and treatment of Alzheimer's disease, including a potential vaccine for the disease.[1] The $50 million for the NIH was prompted directly by the extensive publicity given the first human study of a possible Alzheimer's vaccine developed by scientists of Elan Pharmaceuticals in San Francisco. The results of early phase one studies of a vaccine directed against beta-amyloid fibrils in the brain, with attendant claims of "safety in humans," were announced by the Alzheimer's Association, an advocacy group, at the World Alzheimer Congress 2000 meeting last week in Washington. "It is more clear than ever that the nation must continue its strong bipartisan support for biomedical research on the causes, treatments, and cures for Alzheimer's disease and other diseases affecting millions of Americans," Clinton said. The president, speaking from Camp David, turned his attention from Barak and Arafat to beta-amyloid fibrils. The president said the new vaccine findings "provide new hope not only for Americans who are at risk for developing Alzheimer's disease in the future but for those who are already in its early stages."

No doubt the president's curiosity, like that of millions of television viewers, was piqued by pictures of mice struggling through water mazes. University of Toronto scientists had bred transgenic mice to develop an Alzheimer's-like disease by overexpressing the genes for the beta-amyloid fibrils which drop like sludge in the brains of the sick. They then taught the transgenic mice (PDAPP mice) to swim through a water maze. Over several months, mice vaccinated against the beta-amyloid fibrils remembered how to get through the maze far better than did unvaccinated mice. This was promising evidence that the vaccine may affect ongoing symptoms, and not just prevent the development of new plaques. Histologic data, previously published in *Nature*, had suggested that immunity to beta-amyloid peptide could prevent plaques from forming.[2] And last week Ivan Lieberburg, a coauthor of the study and chief science officer of Elan, reported that the group had obtained evidence that the antibodies had tagged the fibrils causing them to be cleared by microglia, the scavenger cells of the brain.[3]

Former President Reagan's daughter Maureen Reagan, who is a spokesman for the Alzheimer's Association, welcomed the announcement but said much more money is needed; she ventured that an additional $100 million for research would be appropriate for this. "The disease just gets worse every day," she said on CNN's *Late Edition*. When she was asked how the former president is faring, she said, "When I say not so good, Alzheimer's families know what I'm talking about." *Time* magazine also joined the ranks of those crusading against beta-amyloid fibrils and their role in "the aging brain's most heartbreaking disorder." *Time* mourned other celeb victims of Alzheimer's such as Barry Goldwater, Rita Hayworth, and Aaron Copland and suggested that relief might soon be at hand—if only more money were available.[4] Most recently, Gina Kolata, of all people, worried that all this publicity was perhaps premature.[5] Remarkable, since it was Ms. Kolata who had played such a prominent role in the angiostatin hype just a year ago. Anyway, this time she worried that "the eager reporting raises difficult questions of what to say about medical advances and when to say it." It sure does; it also raises questions of what gets funded and why.

If those swimming mice (see below) loosen up a few dollars for science, all to the good. But, I'm afraid that if this bubble of disease-a-week research funding, this medical-crisis-of-the-week mentality, persists, we'll discover nothing much that's really new at all. One week we need to spend billions on AIDS, the next week it's millions for Alzheimer's. Time out, I say: we need money to do medical science where the science itself leads us. If someone were to suggest to President Clinton or his media advisors that he put $50 million into research on familial Mediterranean fever (FMF) or into understanding Portugese Familial Polyneuropathy, they would be laughed out of the pressroom. But, in the first instance, if Mordecai Pras of Israel had not solubilized amyloid fibrils in his search for why patients with FMF get amyloidosis, and in the second, if Dewitt S. Goodman of Columbia University hadn't come across the neurotoxicity of amyloid in his study of genetic variations in vitamin A-binding proteins, the dramatic and exciting work of the Elan scientists wouldn't have been possible.[6,7] I know, because I was around when those observations were made; and no one at the time had Alzheimer's on their mind.

Perhaps, as Lewis Thomas put it, "What [biomedical research] needs is for the air to be made right. If you want a bee to make honey, you do not issue protocols on solar navigation or carbohydrate chemistry, you put him together with other bees . . . and you do what you can to arrange the general environment around the hive. If the air is right, the science will come in its own season, like pure honey.[8]

RANK INJUSTICE

Last week *US News and World Report* issued its annual listing, "America's Best Hospitals."[9] The editors claimed that "The *U.S. News* Honor Roll recognizes hospitals that do many things well, demonstrated by high scores in at least six of the seventeen specialty rankings in 'America's Best Hospitals.' " Hospitals received points based on how high they ranked. Notice that these were seventeen specialties or subspecialties that ranged from "cancer" to "urology"; there was no ranking for "medicine," "surgery" or just plain good doctoring. They obtained these rankings by asking a packet of board-certified specialists about the hospital's reputation, and then added fudge factors like the number of actual/expected deaths, the nurses per patient, the level of "technology," etc. The list came up with the usual suspects in our brave new age of the specialty-oriented, procedure-driven, health care provider (doctor. archaic).

1. Johns Hopkins Hospital, Baltimore (31 points in 16 specialties)
2. Mayo Clinic, Rochester, MN (27 points in 14 specialties)
3. Massachusetts General Hospital, Boston (25 points in 13 specialties)
4. Cleveland Clinic (23 points in 12 specialties)
5. UCLA Medical Center, Los Angeles (21 points in 13 specialties)
6. Duke University Medical Center, Durham, NC (21 points in 12 specialties)
7. Barnes-Jewish Hospital, St. Louis (17 points in 11 specialties)
8. Stanford University Hospital, Stanford, CA (17 points in 11 specialties)
9. Brigham and Women's Hospital, Boston (16 points in 10 specialties)
10. Hospital of the University of Pennsylvania, PA (12 points in 9 specialties)

Who could possibly quarrel with this menu of estimable institutions? No one, on pain of ingratitude for the gifts of medical science to the middle class. Indeed, I would argue that *US News*'s kind of Zagat's survey, a *Michelin Guide* to the comfortable and competent, is a perfect reflection of America's Best Health Care Providers. But, I take a nostalgic exception to this ranking. What's missing from this list are hospitals that mix the philanthropic and the academic in equal proportion. Without the one, the other suffers. What's missing—by my lights— are Cook County in Chicago, Charity in New Orleans, Parkland in Dallas, Kings County in Brooklyn, Bellevue in New York, LA County in Los Angeles, Boston City in Boston. When Boston City was founded in 1860, its mission was spelled out by H. J. Clark and it ought to hold as true today for the Mayo Clinic as for Cook County:

> If poverty is an evil, or disease a misfortune, then certainly, where they coexist, the miseries of each are intensified by the presence of the other, and the sum total is multiplied an hundred fold. . . . It follows that society must still continue to provide for those of its members who are, by either of these calamities, incapacitated from taking care of themselves.[10]

A generation of doctors—and several generations of patients— benefitted from that contract. Nowadays, critics of "paternalistic, hieratic" medicine complain that the patients were warehoused thirty or more to each open ward to serve the role of human guinea pigs. They tell us that the poor were piled bed upon bed like so much industrial inventory for the convenience of doctors, that charity cases provided the fodder of medical teaching. But memoirs of life at Boston City or Bellevue Hospital paint an entirely different picture.[11] They show that the charity patients were the center of a busy hospital life, in which families, friends, clergy, doctors, nurses, medical students, interns, and custodians formed a community. For over a century since 1860, patients in the city hospitals found themselves in precincts that were cleaner, warmer, and more caring by far than any slum in Boston, New York, or Chicago. The records also show that over that same century, doctors and patients were parties to a barter agreement—care was given in exchange for teaching—a largely amicable contract that was unbreached before it was annulled by the HMOs.

Before World War II, the medical services of the great teaching hospitals provided mainly custodial rather than remedial care: food, heat, and a place to get well. The doctors might relieve pain and suffering, but they had few real remedies in hand. They could give arsenic for syphilis, insulin for diabetes, raw liver for pernicious anemia, and antisera for pneumonia. But by and large, as Lewis Thomas recalled, "Whether you survived or not depended on the natural history of the disease.... And yet, everyone, all the professionals, were frantically busy, trying to cope, doing one thing after another, all day and all night."[12] Busiest of all were the interns.

William Peltz, who was to become a professor of psychiatry at the University of Pennsylvania, described the intern's work as "typing blood, doing urinalyses and examining stools, giving transfusions, taking EKGs, typing pneumococci, pronouncing people dead and signing death certificates."[13] The interns were also expected to work in the emergency room and the outpatient department from which they rushed back to the wards to start IVs, perform catheterizations, measure basal metabolism, do spinal taps, place tubes in various orifices, "and more and more."[14] Once past the pup stage, the interns admitted sick people at the rate of four or five per night, obtained their patient's social and medical histories, performed physical examinations, did all but the most difficult laboratory examinations, and after mulling over all other possibilities, committed themselves to the single most likely diagnosis and plan of treatment. Then they waited. Over the next few hours, days, or weeks, they watched—but rarely influenced—the disease until the patient got better or worse; they were again required to keep detailed records of what happened. It was called "keeping"—or later, "buffing"—the chart. Interns were also expected to comfort their patients, to make accurate prognoses—and when all had come to naught—to beg permission from the patient's nearest relative for an autopsy. In the course of these efforts they were expected to work every day and every other night, to ignore weekends—and to remain unmarried.

In return, the interns received room and board and medical training that would last a lifetime. Franz J. Ingelfinger, later the best medical editor of his day, remembered looking up at the stars—or as much as could be seen between the Peabody Building on one side and the house officer's building on the other—and imploring the deities that he might do a good job: "It was an emotional and heady walk between those buildings."[15]

Nowadays, when youngsters who try their hand at industry or politics are called "interns," we tend to forget that the title *interne* derives from the Parisian teaching hospitals of the 1830s. Those institutions, the "Best Hospitals" of France, rewarded their best students with an *internat*. Looking back on the "best hospital" of his own time, Boston City, Lewis Thomas was convinced that the title of intern should be reserved for those who tended the sick *inside* a teaching hospital:

> I am remembering the internship through a haze of time, cluttered by all sorts of memories of other jobs, but I haven't got it wrong nor am I romanticizing the experience. It was, simply, the best of times.[16]

RATS, LICE, AND THOMAS

The best of times wouldn't have happened were it not for Hans Zinsser. Thomas confesses in his memoir that

> I got into Harvard ... by luck and also, I suspect, by pull. Hans Zinsser, the professor of bacteriology, had interned with my father at Roosevelt and had admired my mother, and when I went to Boston to be interviewed in the winter of 1933 [Zinsser] informed me that my father and mother were good friends of his, and if I wanted to come to Harvard he would try to help.[17]

The bacteriologist who had interviewed Thomas in the winter of 1933 and who wished him farewell in 1937 was not simply an eminent professor and an old beau of Grace Peck. When young Thomas first met him, Hans Zinsser was already known as a broadly cultivated polymath, whose reputation in medical research was soon to be matched by one in belles lettres; Zinsser published *Rats, Lice, and History* while Thomas was in medical school (1935). Written as a latter-day variation on Sterne's *The Life and Opinions of Tristram Shandy*, the book gave an entertaining account of how world history has been influenced by epidemics of typhus.

> Infectious disease is one of the few genuine adventures left in the world. The dragons are all dead and the lance grows rusty in the chimney corner. . . . About the only sporting proposition that remains unimpaired by the relentless domestication of a once free-living human species is the war against those ferocious little fellow creatures, which lurk in dark corners and stalk us in the bodies of rats, mice and all kinds of domestic animals; which fly and crawl with the insects, and waylay us in our food and drink and even in our love.[18]

Zinsser followed up *Rats, Lice, and History* with *As I Remember Him: The Biography of R.S.*, a third-person autobiography that survives as a distinguished work of literature. The R.S. of Zinsser's title is an abbreviation of the Romantic Self, or the last letters, inverted, of Hans Zinsser's first and last names. *As I Remember Him* was written two years before Zinsser's death of leukemia in 1940 at the age of sixty-one. A selection of the Book of the Month Club, it had reached the best-seller list as its author lay dying; news of its warm reception by book reviewers filtered into the obituaries. A worn copy was another of those books on Thomas's office shelf. The volume was widely popular among doctors. Zinsser entered Columbia's College of Physicians and Surgeons in the class of 1903, a year ahead of his friend-to-be, Joseph Thomas, Lewis's father. Zinsser and Joseph Thomas were successful enough in medical school for both to earn internships at Roosevelt Hospital, where Zinsser acted as a cicerone for Joseph. In 1915 Zinsser accompanied the American Red Cross Sanitary Commission to investigate a devastating outbreak of epidemic typhus in Siberia. After much trial and error, he succeeded in isolating a variant form of the microbe that caused typhus and developing a vaccine against it. His scientific gifts were not limited to microbe hunting, however. He also became a prolific medical writer and editor: his standard *Textbook of Bacteriology* went through many editions. After faculty positions at Stanford and Columbia, he was appointed to the chair at Harvard in 1923. And there, his research, which had started out as a field study of how *rickettsiae* cause specific forms of typhus, moved rapidly into the young science of immunology. He belonged to the Association of American Immunologists, the National Academy of Sciences and, not coincidentally, the American Commission for the Control of Rheumatism.

Like Lewis Thomas, Zinsser died of a hematologic malignancy. The last chapter of *As I Remember Him* forecasts in stoic detail the events of Zinsser's terminal illness. He writes of himself in the third person:

As his disease caught up with him, RS felt increasingly grateful for the fact that death was coming to him with due warning, and gradually. So many times in his active life he had been near sudden death by accident, violence, or acute disease.... But now he was thankful that he had time to compose his spirit, and to spend a last year in affectionate and actually merry association with those dear to him.[19]

July 13, 2000

AIDS in Africa;
Gene Death in America

DENYING THE HOLOCAUST

OPENING the 13th International AIDS Conference in Durban, South African president Thabo Mbeki disappointed his large audience—and many more listening at home—by not retracting his perverse notion that AIDS was not caused by HIV infection. Although 4.2 million people, or 20 percent of South Africa's adult population, is infected by HIV, as compared with 3 percent in 1993, Mbeki maintained that diseases such as poverty, vitamin A deficiency, cholera, syphilis, and "other complicated Latin names," not HIV, were responsible for Africa's woes.[1] Mbeki has been consulting such deniers of the HIV-AIDS link as Peter Duesberg and David Rasnick while vigorously arguing for "an African solution" to the AIDS problem.[2] This sort of rhetoric sounds very much like the argument for a "German physics" or a "Russian genetics," which led to Heisenberg's miscalculations on the one hand and to Lysenko's crop failures on the other.

These are the facts: HIV causes AIDS, which is spread by live, virus-containing cells that display well-defined cell surface receptors for the virus. No receptors, no viral infection. The virus-infected cells are present in male or female sexual discharges. The rectal and oral mucosa are somewhat more permissive to viral exchange than is the vagina; trauma of any sort facilitates infection and condoms are partially protective. If one prevents penises from entering various orifices, one can decrease the incidence of the disease. These facts are difficult to face if social constraints or private urges persuade one group or another to deny them. Duesberg and his colleagues are wrong. Babies catch the same HIV virus subtype that their mothers have, and if you kill that virus subtype with specific drugs, the disease goes away. Remembering Koch's postulates, which taught us how syphilis or anthrax infect their victims, will do us all some good.

Until now, however, "African solutions" seem not to have worked. Despite the efforts of enlightened, hardworking activists who are trying their best to stem an epidemic that has already infected 24.2 million people, human survival itself seems at risk in sub-Saharan Africa. Karen Stanecky of the U.S. Census Bureau reported to the conference in Durban that in Botswana life expectancy is now thirty-nine, instead of the seventy-one enjoyed by the developed world. She estimated that by 2010 many countries in southern Africa will see life expectancies falling to near thirty.[3] David Ho of Rockefeller University's Aaron Diamond AIDS Research Center had the proper response to Mbeki on July 11: "The failure to properly address the modern plague caused by HIV is an act of (irresponsibility) that will be judged by history."[4]

The response of many other African leaders and health officials has been more responsible. They demand cheap or free drugs against AIDS along with help from the industrialized world in educating a population the folkways of which are, to say the least, traditional and patriarchal. In days to come we'll read news of this grant à la the Gates Foundation, or that donation à la Merck. They're certainly well intentioned, but I'm afraid it's spitting in the ocean. The only solution for the pandemic of AIDS is a cheap vaccine à la Sabin or Salk. Global enlightenment wouldn't hurt, either.

MEDICAL ERRORS

It's good to know that AIDS is not as great a threat in the U.S. as it is in Africa. Our public health is in greater danger from another menace: medical error. That's right; doctors, not viruses, are this nation's major culprit. Medical mistakes kill between forty-four thousand and ninety-eight thousand people a year, in hospitals alone. According to a panic-inducing study released last fall by the National Institute of Medicine, more people are killed every year by medical error (read: doctors) than die of car accidents (43,458), breast cancer (42,297), or AIDS (16,518).[5]

Based mostly on meta-analysis of published studies and chart reviews, the Institute of Medicine of the National Academy of Science (IOM) report, supervised by HMO types, presented its data in the most dramatic fashion possible. The report begins: "The knowledgeable health reporter for the *Boston Globe*, Betsy Lehman, died from an overdose during chemotherapy. Willie King had the wrong leg amputated . . ." President Clinton went into action and proposed a new office to regulate medical error, Congressional action was planned, and press release followed press release. Not unexpectedly, the two major proponents of the study, Drs. Lucian Leape and Troyen A. Brennan, both at Harvard, went on the road and hit the journals with their message of revolutionary action against killer doctors.[6]

But this spring something remarkable happened on the way to the tumbrels. Brennan had a change of heart. In a *New England Journal of Medicine* article he suddenly confessed that "neither study cited by the IOM as the source of data on the incidence of injuries due to medical care involved judgments by the physicians reviewing medical records about whether the injuries were caused by errors.[7] Indeed, there is no evidence that such judgments can be made reliably."

It turns out that what the report didn't tell us is that the death rates for AIDS, breast cancer, and car accidents were taken from actual death certificates, while the figures for deaths due to medical error were extrapolated from two retrospective chart reviews (one in Colorado and Utah; the other in New York State) of relatively small populations. This week researchers from the University of Indiana led by C. J. McDonald took another look at these data and threw cold water on the IOM fire.[8] The McDonald brigade arrived just in time to douse a Washington ablaze, thanks to the inflammatory tone of the IOM report.

The McDonald response in *JAMA* was devastating. Based on thorough statistical analysis, the Indianapolis group concluded that "the available data do not support IOM's claim of large numbers of deaths caused by adverse events, preventable or otherwise." As might be expected, Leape is still hanging in there. Even if the hospital-based data were soft, he argued, there are more accidents out there that were not reported, since more than half of surgeries are now performed on an outpatient basis. I find this retort lame and conclude from my own review of the data that the numbers are inconclusive. Of course medical errors occur and only rigorous training and care can be expected to prevent them. Incompetents creep in, and sharp eyes need to look out. Big news.

July 4, 2000

Patrick O'Brian: The British Navy
and the Human Phenome Project

IT S BEEN a long holiday weekend by the seaside here at Woods Hole, and I've just finished devouring Patrick O'Brian's last book, *Blue at the Mizzen*.[1] Literary folks tend to dismiss O'Brian as a minor writer of maritime adventures, a chap who wrote Harry Potter yarns for sea-loving geezers. Well, perhaps there's something to be said for that view, but to me he's one of the few writers who has managed to get both Darwin and the Royal Navy right.

AUBREY AND MATURIN

Blue at the Mizzen is the twentieth in a series of novels that begins with an acrimonious encounter between Lt. Jack Aubrey RN and Dr. Stephen Maturin during the interval of the first and second movements of Locatelli's C major quartet in the music room of Government House, in Port Mahón, Minorca, on April 1, 1800. They become improbable friends after Aubrey appoints Maturin as surgeon to his ship. Soon the game is afoot, a phrase from another unlikely pair of chums, Holmes and Watson. There is not a minute to lose, as Aubrey would have it, and they are off to fight Napoleon on the seven seas. For the next fifteen years their brotherhood grows in the course of what the *New York Times* celebrated as a "struggle against the perfidy of man and the unpredictability of nature." The books are tales of "Bankruptcy, betrayal, human weakness, unrequited love, imprisonment, disease, shipwrecks and storms . . ."[2]

Indeed they are, but O'Brian has not simply written a nineteenth century romance of two gents on the high seas in high dudgeon, an adventure yarn à la Conan Doyle. Nor are his heroes cast in the mold of Holmes and Watson. On the contrary, O'Brian's epic describes how tough it is to maintain a global empire, the job of the Royal Navy, and how hard it is to describe the globe, the goal of the the Royal Society. O'Brian turns this dual narrative into a major work of art because his fiction transcends genre and his prose is in perfect pitch. I'd argue that he has written neither nautical nor science fiction, but literature.

When O'Brian died last January, he missed the snickering that erupted among literati when Dean King published his biography of O'Brian in March: The story of O'Brian's life—or at least as told by King—reads more like John Le Carré than Patrick O'Brian.[3] It had always been assumed—indeed it had been mooted by the author himself—that he was a born and bred Irish Catholic; in fact, O'Brian spent his winters in Trinity College, Dublin, and lived for many in Collioure, a French town bordering Catalonia. Actually, he was born Richard Patrick Russ, grandson of Carl Gottfried Russ, a Jewish furrier who arrived in London in 1860 to found a prosperous firm on New Bond Street. His father, Dr. Charles Russ, was a physician of questionable probity (he treated gonorrhea by electrolysis) who raised him as an Anglican. O'Brian was born on December 12, 1914, as the last of eight children in Chalfont-Saint-Peter, Buckinghamshire, and educated at a minor public school in Torrington, Devon. He never attended university. In 1936, O'Brian married his wife, Sarah, and a year later fathered a son, Richard, who still survives him. In 1939 Sarah had a daughter, Jane, who suffered from spina bifida and died at the age of three. Almost immediately after the child's death, O'Brian left his wife. During the war, O'Brian drove ambulances early in the Blitz, but spent the rest of the war with the Special Operations Executive, where he ostensibly engaged in "political warfare" at a desk. At about this time, he changed his name to O'Brian, ostensibly to conform with that of a brother who needed a nom de guerre in order to "enlist in the air force" (tell that to the marines). O'Brian married again in 1945, to Frieda Mary Wicksteed (the mother of Count Nikolai Tolstoy by an earlier marriage), with whom he may or may not have worked in military intelligence. Their son, Count Nikolai Tolstoy-Miroslava, became a well-respected historian. O'Brian's works—in addition to his twenty-volume epic—include biographies of Picasso and Sir Joseph Banks, several children's books, translations of Simone de Beauvoir, Jean Lacouture, and Colette, and an early novel *Testimonies*, a work that Delmore Schwartz reviewed as "inspired."[4] This isn't Conan Doyle.

DARWIN AND FITZROY

Literature aside, I'm going to call attention to another aspect of the O'Brian oeuvre. It has long seemed to me that O'Brian's two heroes are based in good part on the lives of Capt. Richard Fitzroy of the Royal Navy and Charles Darwin of the Royal Society. Before Darwin could come up with his theory of evolution, a century and a half of natural history was required to provide a map of what I would call the human phenome (phenome, as in "phenotype"). It was not until this year that the human genome project spelled out the order of genes in our chromosomes. It was not until the middle of the nineteenth century that naturalists such as Banks or Cuvier finished enumerating the human phenome—a catalogue of the diversity of life on this planet, otherwise known as the Great Chain of Being. Without this hierarchical numbering of the species by zoologists we wouldn't have had *The Origin of the Species*. Without comparing the anatomy of all those species to man, we wouldn't have *The Descent of Man*. Appropriately, O'Brian awards fellowships in the Royal Society to both Aubrey (as meteorologist) and to Dr. Maturin (as comparative anatomist). Morever, the books are chock-a-block not only with a lexicon of nautical jargon but also with minutiae of eighteenth century cuppings, bleedings, trephinings, purgings, and the greatest catalogue of wildlife and anatomical curiosities since Cuvier. Oh yes, Maturin knows the author of *La Regne Animal* well; in this passage he examines the library of his chief at the admiralty, Sir Joseph Blaine, FRS, a distinguished entomologist:

> After a while Stephen turned to the bookshelves: Malpighi, Swammerdam, Ray, Réaumur, Brisson, the most recent Frenchmen, including the elder Cuvier's latest essay, which he had not yet seen. He read the first chapters, sitting on the arm of his chair, and then moved over to Sir Joseph's cabinet to find the insect in question. Drawer after drawer filled with creatures, lovingly killed, pinned down and labeled: in the second drawer he saw that great rarity, a true gynandromorph, a Clouded Yellow, male one side, female the other . . . [5]

O'Brian confesses in one or another of his prefaces that his detailed accounts of naval action and political intrigue are but slightly rearranged from Royal Navy documents and letters of the day, but remains somewhat reticent on the Darwin/Fitzroy link. Nevertheless, when one compares the itinerary of HMS *Beagle* in 1836 (the Cape Verde Islands, the South American coast, the Strait of Magellan, the Galapagos Islands, Tahiti, New Zealand, Australia, the Maldives, and a stop at Mauritius before the home leg to England) with the voyages of Aubrey and Maturin, the itineraries come uncommonly close. Like O'Brian's fictional Aubrey, the real Fitzroy was a scion of country gentry, a member of parliament, and earned his FRS in 1851 for inventing the modern weather forecast (by means of the Fitzroy barometer).

In his other life, O'Brian has written a biography of Sir Joseph Banks, FRS, the model for Sir Joseph (*vide supra*). After Banks gained fame as the naturalist on Captain Cook's voyages, he worked in the topgallants of the admiralty, became the doyen of beetle experts, and was elected president of the Royal Society. Add Banks to Fitzroy and Darwin (Blaine to Aubrey and Maturin) and we have an epic à clef. The key is given away when Stephen Maturin and his mate beg Jack

Aubrey to let them land on the Galapagos Islands: "How I yearn to set foot on one of those islands. Such discoveries in every realm." Aubrey replies: "Stephen, here I have my tide, my current and my wind all combined—my enemy with a fine head-start so that here is not a moment to be lost—could I conscientiously delay for the sake of an iguano or a beetle—interesting, no doubt, but of no immediate application in warfare? Candidly, now?" Stephen replies: "Banks was taken to Otaheite to observe the transit of Venus, which had no immediate practical application." "You forget that Banks paid for the Endeavour, and that we did not happen to be engaged in a war at the time: the Endeavour was not in pursuit of anything but knowledge."[6]

Read in the context of the major scientific discovery of the nineteenth century (Darwin's), O'Brian's work is a tribute to the homage that science owes to military power. The century-long, hot and cold war between England and France for global dominion was accompanied by the flowering of Western science. When Danton urged the *convention* to action in 1792: "Il nous fait l'audace, encore de l'audace, tojours l'audace," he was echoing the temper of his time—and of Aubrey's: No *l'audace*, no ocean exploration. Indeed, it could be argued that the hot and cold war between the Western democracies and the totalitarian states played an equivalent role in the twentieth century. No *l'audace*, no exploration of outer space—nor of our genes. This larger worldview is the kind of discourse that O'Brian's sea stories consciously elicit.

ELIOT AND AUDUBON

While O'Brian wished very much to wind up on the library shelf within a foot or two of Jane Austen, his favorite author, there are overtones in his work of other masters. His social urges are closer to those of George Eliot (his take on the enclosure movement in *The Hundred Days* is that of Eliot's *Felix Holt*) and his lyric attention to the natural world is worthy of Audubon's *Ornithological Biography*. O'Brian echoes Audubon when Maturin finds a fine great company of snowy white egrets accompanied in flight by "a glossy ibis, absurdly black in this light and company, and continually uttering a discontented cry, something between a croak and a quack." Stephen "had the impression that the ibis was extremely indignant at the egrets' conduct: and indeed so late a migration, well on in the month of May, was unusual, unwise, against all established custom."[7] That isn't just natural history, it's natural writing.

When Patrick O' Brian died this January, fine eulogies issued from every corner of the world of letters. William F. Buckley hailed him as "a naturalist, linguist, translator, biographer, the most evocative writer on the sea since Homer."[8] David Mamet, who claims to have read all twenty of the Aubrey/Maturin novels three or four times, compared the heroes of O'Brian's epic to Holmes and Watson "blessed in having, in my generation, an equally thrilling set of heroes with characters that have become a part of my life."[9] George Will praised him for developing the themes of "courage, honor and gentle manliness" in episodes that "constitute a single 6,443-page novel, one that should have been on those lists of the greatest novels of the 20th century."[10] O'Brian would not have been surprised. Shortly before his death he had completed the last of his triumphant book tours in the United States where his fans, which included such a mixed lot as Joan Didion, Walter Cronkite, Charlton Heston, and Robert Hass (the U.S. poet laureate), had been his chaperones. He appeared before socko crowds at Stanford, Rockefeller University, the New York Public Library—and the model room of New York Yacht Club. He deserved all of it.

June 26, 2000

The Human Genome
Is (Almost) Complete

MEN ON THE MOON

THIS WEEK one medical story dominates all others. On Monday, June 26, 2000, scientists announced that the human genome had been largely deciphered. It is at once a splendid product of the human mind and a tribute to the reductionist aims of modern science. By my lights, it's one great step for mankind.

We owe this achievement to two rival groups: a worldwide consortium of sixteen university laboratories led in the United States by Francis Collins of the NIH/DOE, and to a private company, Celera Genomics of Rockville, MD, J. Craig Venter CEO.[1,2] At news conferences that erupted from London to Washington, it was announced that the consortium had mapped 97 percent of the human genome and sequenced in detail 85 percent of the approximately 3.2 billion base pairs of DNA. Celera announced Monday that it has completed 99 percent of the genetic sequence, albeit with small gaps. But the general outline is there and both groups are arranging for full publication in *Science* this fall.

Because of the time difference, the British and French members of the consortium had hit the wires first.[3] Dr. Michael Dexter, of the Wellcome Trust, which committed £210 million to the British effort, said immodestly that the first phase of the work marked a medical landmark: "Mapping the human genome has been compared with putting a man on the moon, but I believe it is more than that. This is the outstanding achievement not only of our lifetime, but in terms of human history."

I was reminded of a plea by Lewis Thomas, first delivered before a congressional committee in the 1980s, when the genome project was first being bandied about:

> If I were a policy-maker, interested in saving money for health care over the long haul, I would regard it as an act of high prudence to give high priority to a lot more research in biologic science. This is the only way to get the full mileage that biology owes to the science of medicine, even though it seems, as used to be said when the phrase still had some meaning, like asking for the moon.[4]

The "outstanding achievement of our lifetime" was climaxed by a White House press conference at which Collins and Venter gave cheek-to-jowl versions of how far each team had come, what the discovery meant for human health and why this was a great day. "Today we are learning the language in which God created life," said the president, appealing to the deity with an election year at hand and polls telling the White House that half of America thought that the genome project was basically immoral.[5]

The meeting was on Celera's timetable, not the consortium's. Venter entered the race in 1998, a decade after James Watson, the first director of the NIH Human Genome Project, had predicted that it would be finished in 2005 at a cost of $2 billion. Venter had it done in two years at a cost of $250 million, thereby both winning the race by a mile and justifying Celera's brisk motto: Speed Matters. He had also vastly accelerated the consortium's effort. This spring, Collins, who became Watson's successor at the NIH, was in the position Neil Armstrong might have been had he landed on the moon to find Lee Iacocca of Chrysler waiting there to greet him.

Irony of ironies: the president congratulated James Watson—a guest at the conference—for his foresight by quoting from the classic Watson-Crick paper of 1958 in *Nature*.[6] Clinton did not call attention to Watson's own account in *The Double Helix* of his delight in beating out Rosalind Franklin and Linus Pauling in the race for the structure of DNA. Even then, Speed Mattered. At times during the conference at the White House, Collins and Venter looked as comfortable with each other as Barak and Arafat. Indeed, over the last year, acrimonious differences between the two groups had surfaced in the press and it became obvious that the two protagonists are of dissimilar temperament. Collins is a serious, devout Christian who professes his fundamentalist faith in a designed universe. "We have caught the first glimpses of our instruction book, previously known only to God," said Collins at the White House, looking to heaven.[7] Venter, on the other hand, is an earthy entrepreneur with a mission to accomplish, a business to build, and a drive for practical results. Venter spoke of seeing people die in Vietnam, where he served as a medic, and paid simple homage to the human spirit.[8]

James Watson was moved to kinetic, informational rhapsodies: "Now we have the instruction book for human life." Watson told the BBC, "Things are just going to move faster. After the printing press, there was an explosion, more people could have information. We'll understand ourselves better, have a better idea of what human nature is."[9]

THOUGHT'S NEWFOUND PATH

Watson was right, of course: there was indeed an explosion after Gutenberg's discovery of movable type. In the United States there was an even greater explosion after Morse's telegraph in 1840 and the laying of the transatlantic cable. I was reminded of that nineteenth century achievement as I watched Tony Blair's jejune close-up over the satellite screen in back of Clinton's head (the prime minister was there to pull the British oar on behalf of the consortium). The public/private argument with respect to technologic advance was as great an issue in 1858 as it is today. Ralph Waldo Emerson, the transcendental sage, was on a canoe trip in the Adirondacks in 1858 when he heard the news of the transatlantic cable:[10]

> Of the wire-cable laid beneath the sea,
> And landed on our coast, and pulsating
> With ductile fire. Loud, exulting cries
> From boat to boat, and to the echoes round,
> Greet the glad miracle . . .

Emerson's companions on the trip were a group of Harvard literati and "philosophers" including Louis Agassiz; Dr. Jeffries Wyman, an eminent anatomist; James Russell Lowell of the *Atlantic Monthly*; and Dr. Oliver Wendell Holmes's brother John. Emerson's response to the news of technical success in the nineteenth century was much like Jim Watson's response in the twenty-first: he too was glad that "people could have more information," could have "a better idea of what human nature is." Emerson, too, was worried that a profit-making corporation had done that which the government, or its universities, should have accomplished. He was well aware of the claim by a Harvard physician, Dr. Samuel Jackson, that it was he who had proposed the telegraph to Morse. Emerson was disappointed that:

> . . . a hungry company
> Of traders, led by corporate sons of trade,
> perversely borrowing from the shop the tools
> Of science, not from the philosophers,
> Had won the brightest laurel of all time.

In 1858 as in 2000, private industry had beaten out the universities in the race for the "brightest laurels of all time."

Nevertheless, to a large degree, the tools of science are still forged in our universities. Venter may have spurred the work of consortium, but the opposite is also true. Without public funding of the basic science and immense treasure-house of data that Collins and his team have posted on the web, Venter would still be at it. Emerson might be the first to salute not only Watson and Collins, but Venter as well. He would thank the whole consortium and the troops of Celera for making possible that

> Thought's new-found path
> Shall supplement henceforth all trodden ways,
> Match God's equator with a zone of art,
> And lift man's public action to a height
> Worthy the enormous cloud of witnesses.

June 13, 2000

Herbs, Genes, and Abortion

THIS WEEK news was made by a Chinese herb in Brussels, a French drug in Washington, and an American editor in Florence. Each story contains a touch of politics, a brush with science, and a lesson for the clinic.

HERBAL TROUBLE

In the latest issue of the *NEJM*, Belgian clinicians report on their experience with "Chinese herb nephropathy."[1] The herb, *Aristolochia fangchi*, had been a constituent of weight-loss pills prescribed in Belgium from 1990 to 1992, when a semi-epidemic of the kidney disease led Belgium to ban the herb. Since then, the Hôpital Erasme in Brussels has treated 105 patients for severe kidney damage. Of 43 patients with end-stage Chinese herb nephropathy who were being treated with either transplantation or dialysis, 39 agreed to undergo prophylactic surgery. The authors found 17 cases of carcinoma of the ureter, renal pelvis, or both; they also found one papillary bladder tumor. Nineteen of the remaining patients had mild-to-moderate urothelial dysplasia, and two had normal urothelium. All tissue samples analyzed contained aristolochic acid-related DNA adducts (the biochemical hall-mark of herbal injury). The cumulative dose of *Aristolochia* was a significant risk factor for urothelial carcinoma, with total doses of more than 200 g associated with a higher risk of urothelial carcinoma.

In his *NEJM* paper, Dr. Nortier gently concludes that "Our findings reinforce the idea that the use of natural herbal medicine may not be without risk." But, on the editorial page, Dr. David Kessler, former head of the FDA and now dean of Yale's medical school, sounds a sterner note.[2] He complains that by passing the permissive 1994 Dietary Supplement Act, "Congress has put the FDA in the position of being able to act only after the fact and after substantial harm has already occurred." Pointing out that it took a tragedy (the toxicity of sulfonamides) for Congress to pass the Food, Drug, and Cosmetic Act of 1938, the dean argues that "examples like that described by Nortier et al. should persuade Congress to change the law to ensure the safety and efficacy of dietary supplements before more people are harmed."

Acting after the fact, and after Kessler's version of Émile Zola's *J'Accuse*, the FDA announced Friday that it is stopping importation of herbs in the Aristolochia family. The agency advised health professionals that the ingredient had been present in the Belgian preparation because *Stephania tetrandra*, a botanical not known to contain aristolochic acid, had been inadvertently substituted with the botanical *Aristolochia fangchi*, which contains aristolochic acid as a normal constituent "because of the similarity of the Chinese names for these 2 botanicals."[3]

I would add that while patients have some control over whether they will take herbal medicines for weight reduction, they are less constrained in the case of painful diseases like rheumatoid arthritis or recurrent kidney stones. Herbal practitioners—and some pharmaceutical enthusiasts—prescribe thunder god vine for the joints and madder roots for the stone. Both possess significant, well-described, toxicities: the vine causes aplastic anemia and the root is carcinogenic.[4] But the FDA is silent on these botanicals and the NIH is sponsoring work on thunder god vine.

It looks as if there are battles enough in the herbal trenches for an army of Kesslers—or perhaps an armée of Zolas, who, after demolishing the opposition by sharp reason, moderated the temper of his *J'Accuse* in his peroration: "As for those I have accused, I do not know them, I never saw them, I have against them neither resentment nor hatred. They are for me only impersonal forces, spirits of social mischief. And my complaint here is only that of an ordinary dissident who wishes to hasten the triumph of truth and justice." ["Quant aux gens que j'accuse, je ne les connais pas, je ne les ai jamais vus, je n'ai contre eux ni rancune ni haine. Ils ne sont pour moi que des entités, des esprits de malfaisance sociale. Et l'acte que j'accomplis ici n'est qu'un moyen révolutionnaire pour hâter l'explosion de la vérité et de la justice."][5] Right on Dr. Kessler.

RU 486: GUADELOUPE V. USA

Last Wednesday, Gloria Feldt, the head of Planned Parenthood, announced that unduly restrictive proposals by the FDA might limit access of the early-abortion pill RU 486 (mifepristone) to women who need it most.[6] The Feds have been dawdling in their approval of the drug since last February when the FDA promised to approve its sale, provided that certain final, but undisclosed, requirements were met. Ms. Feldt said that the nonprofit Population Council of New York, sponsor of the pill, told her group that the FDA is proposing needless curbs on the use of mifepristone.

"We are deeply concerned that the FDA is considering restrictions that in my view would virtually assure that very few doctors would ever make mifepristone available," Feldt said last week. She was chiefly concerned that the FDA proposed that physicians who were to administer RU 486 must be part of a registry, which she said would deter doctors worried about anti-abortion violence from offering the pill. Similar worries dictated that it was the philanthropic Population Council, rather than a large pharmaceutical house, which in 1994 took over the U.S. marketing rights to RU 486 from the original French manufacturer.

Feldt said the FDA also was considering "long-term health tracking" of at least some RU 486 recipients, something she called unnecessary because half a million European women have used the pill successfully since 1988.

Not only European women are ahead of their U.S. sisters in this regard. A recent meta-analysis of worldwide data by University of California, San Francisco scientists showed that in fifty-four studies the combination of mifepristone and misoprostol (a prostaglandin E1 analogue) is 95.5 percent effective in causing abortion when taken within the first forty-nine days. Complications were very minor in this early abortion population.[7]

More convincing still is a recent report from a society perhaps less advanced medically, if not socially, than the Bay Area.[8] The authors reported that "shortly after mifepristone's introduction in Guadeloupe, a semi-developed Caribbean territory administered by France, in 1991, two of the authors conducted a small prospective study of a one treatment-visit regimen. The study regimen was subsequently adopted as the standard of care for medical abortion on the island." In Guadeloupe, the success rate (95.4 percent) was comparable to rates found in metropolitan France and the authors were proud to announce that "Protocol adherence appeared to be excellent and loss to follow-up was rare. We suggest that home administration of misoprostol can be safe and effective in most nonindustrialized settings."

Back in the U.S., the FDA declined to respond to the statement by Planned Parenthood. Officials are probably considering the safety and efficacy of RU 486 in the setting of an industrialized nation prone to anti-abortion violence.

DRAZEN IN FLORENCE

Jeffrey M. Drazen, a distinguished physician/scientist who directs the pulmonary division at Brigham and Women's Hospital in Boston, has recently been appointed the new editor of the *New England Journal of Medicine*. He also happens to be working in an area of research that overlaps some of my own interests, and it was therefore with some pleasure that I went to hear his plenary lecture at the 11th International Conference on "Advances in Prostaglandin and Leukotriene Research" held in Florence, Italy, from June 4–8. Drazen's paper was the opening round in the new field of pharmacogenomics. That daunting neologism describes a discipline that tries to decipher the variations in our genes that determine how we respond to drugs.

It has been very clear for some time that the same drug will affect different individuals in very different ways, and that such differences are genetic. But now that we have the genes spelled out exactly, we can start to study populations of patients that do or do not respond to new drugs. It has also been clear, from the work of Drazen's long-term collaborator and mentor, K. Frank Austen, that cysteinyl leukotrienes (Cys LT) formed via the 5-lipoxygenase (5-Lo) pathway mediate some—if not all—the events of bronchial asthma.

Drazen has summarized the situation: "These agents either prevent the synthesis of the leukotrienes, by preventing the action of the 5-lipoxygenase-activating protein or the catalytic action of the 5-lipoxygenase, or by inhibiting the action of leukotrienes at the CysLT receptor. Numerous clinical trials in exercise-induced asthma, allergen-induced asthma, aspirin-induced asthma, and spontaneously occurring asthmatic episodes have indicated that these agents are safe and effective asthma treatments."[9]

But not all patients respond to such leukotriene modifiers in the same way. In Florence Drazen reported that in the 5-lipoxygenase pathway there is a family of polymorphisms in the core promoter of the 5-LO gene which consists of the "addition or deletion of Sp-1/Egr-1 DNA binding motifs." The molecular jargon simply means that if you read the text of DNA with your finger, you either run your fingertip smoothly across the paper or ruffle the page by pressing too hard.[10]

Drazen found that variants in these sequences were associated with diminished gene transcription of the 5-lipoxygenase, and that patients who had no wild-type sequences to compensate for these variants had diminished responses to anti-5-LO agents. These data, collected from a large clinical sample of patients who had undergone phase three testing with a leukotriene modifier, montelukast, showed convincingly that the genetic program of nonresponders differed from the responders at these exact loci.[11]

The audience, which included many of those who had first described and isolated prostaglandins and leukotrienes and who had first spelled out their role in disease, appreciated this fine extension of a long, common effort. They also appreciated that the new editor had announced a new chapter in the book of genes.

May 23, 2000

Chronic Fatigue and
the Wisdom of the Body

THIS WEEK the news comes from the fields of inflammation and the imagination. Tales of a fevered imagination bear the mark of Edgar Allan Poe, but it was Walter B. Cannon, a student of William James, who taught us that our imagination can cloud the wisdom of the body to produce what we now call chronic fatigue syndrome.

EATING THE DEAD

Cells die from many causes, and when they die they often provoke inflammation. That's no great surprise; think of your last twisted ankle or the state of your fingers after the car door slams on them. That sort of cell death is called necrosis and is followed by an acute inflammatory response. On the other hand, cells die throughout our lifetime as our bodies grow and change. Menstruation in women and baldness in men are caused by death of uterine and follicular cells, respectively; by and large they proceed in the absence of inflammation. This type of cell death, which takes place as tissues are remodelled all over the body, is called apoptosis. The mechanisms of apoptosis, and its distinction from necrosis, have become critical to fields as far apart as cardiology and oncology.[1]

Peter Henson and his colleagues in Denver have figured out how we manage to dispose of dead cells without becoming inflamed.[2] They argue that the endpoint of apoptosis in life is the disposal of cellular corpses by professional phagocytes (phagos = eat, cytos = cells), a term introduced by Metchnikoff a century ago. But these cells, such as macrophages, ought to release mediators of inflammation. Why don't they? The answer lies in how the lipids of our cell membranes are assembled.

During apoptosis, the asymmetry of plasma membrane phospholipids is lost, which exposes one of these lipids, phosphatidylserine (normally on the inner leaflet of the cell membrane), to the external surface. The phagocytosis of apoptotic cells can be inhibited stereospecifically by phosphatidylserine or its structural analogues, but not by similar phospholipids, suggesting that phosphatidylserine is specifically recognized. Henson et al. reasoned that there must be a receptor on macrophages for dead cells, and that receptor must be one for phosphatidylserine.

This week, Henson and his colleagues reported that they have cloned the gene for the phosphatidylserine receptor, transfected it into B and T lymphocytes, and permitted these nonphagocytic cells to recognize and engulf apoptotic cells in a phosphatidylserine-specific manner. Using a monoclonal antibody against this new receptor they not only found it on the surface of macrophages, fibroblasts, and epithelial cells but showed that the antibody inhibited the phagocytosis of apoptotic cells. Finally, they came up with the ace in the hole of apoptosis. When macrophages take up apoptotic cells they release TGF beta (an anti-inflammatory cytokine) and turn off their production of TNF alpha (the major pro-inflammatory cytokine). The opposite is true when cells take up necrotic material or debris tagged by antibody and/or complement. Sure enough, their monoclonal antibody, signalling via the "dead cell" receptor, induced an anti-inflammatory state in macrophages. So now we know why apoptosis doesn't cause inflammation: the "dead cell" receptor sends signals that end the wake.

A HOLE IN THE HEAD

If you could actually yank the control box from one of Antrin's busily cerebral novels, God knows what you'd find there. Bats maybe. An audiocassete edition of the Encyclopedia Britannica, recited by Monty Python. And probably the shrunken heads of Thomas Pynchon, Nicholson Baker, Edgar Allan Poe and Donald Barthelme, writers whose DNA seems to have bled into Antrin's own.[3]

There are many of us who are a bit skeptical about the whole spectrum of diagnoses that fall under the rubrics of "chronic fatigue syndrome," "myalgic encephalitis," "irritable bowel syndrome," "total chemical allergy," etc. There is no question but that the patients suffer—and often terribly—from these conditions, and again no question but that their disability is real. Our skepticism arises from the fact that the hallmark of each disease is that its diagnosis requires the complete absence of objective physical or biochemical derangement. We wonder whether these patients are not really the victims of a complex set of socially and medically constructed diseases—such as the "railway spine," "chronic appendicitis," or "female hysteria" favored by nineteenth century clinicians. These doubts are summarized in Edward Shorter's *From Paralysis to Fatigue: A History of Psychosomatic Illness in the Modern Era*.[4]

One's doubts are reinforced by work reported in the most recent issue of the *Journal of the Royal Society of Medicine*.[5] Arguing that "little has been reported on prognostic indicators in children with chronic fatigue syndrome (CFS)," they used interviews with children and parents, a mean of 45.5 months after the onset of illness, to follow up twenty-five cases of CFS referred to tertiary pediatric psychiatric clinics. In keeping with the anecdotal nature of the field, the authors fail to discuss such simple criteria of ordinary medicine as physical signs or laboratory data.

But they found real evidence that this syndrome, which one might call "Munchhausen's fatigue by proxy," seemed to begin in the autumn school term: children were most likely to develop CFS in the autumn term when they start secondary school. Seventy-six percent of children developed CFS between September and December. On average, the children were eleven years old when the illness began, coinciding with the move to secondary school. They weren't suffering from any known psychiatric illness, either, as a companion study showed.

Infective diseases in childhood are far more common in kids from poor families. It was noteworthy that this study of childhood CSF supported earlier data on CSF in adults that showed most of its sufferers were from "higher socioeconomic" (i.e., rich) families. "This probably reflected the greater medical resources available to parents belonging to higher social classes," the authors told Reuters.[6] One might point out that the direct correlation between income and CSF might argue for the social-construction hypothesis versus the usual "infective" or "somatic" etiology of these troubling conditions. Shorter's book sheds light on the history of these socially constructed conditions. He properly argues that patients who subscribe to the diseases of the Zeitgeist really *do* suffer from their illness—whether in body or mind or both—and therefore deserve not only the attention but compassion of doctors. He urges doctors not to regard "patients with 'somatoform' symptoms as bizarre objects but as individuals who enjoy the dignity that all disease confers; our task is rather to understand why the kinds of psychosomatic symptoms that patients perceive change so much over the ages."

I would add that our understanding of why the symptom pool has shifted from the motor side of the nervous system (hysterical paralysis à la Charcot or Freud) to the sensory side (chronic fatigue syndrome à la Jane Brody) is a history of sensibility rather than a history of events. The construction of these symptom complexes has, after all, been the work of *men* figuring out what is wrong with *women*. This transformation, I would guess, reflects in part the changing role of women over the last century: symptoms of motor paralysis in the days of female passivity have been replaced by symptoms of sensory fatigue in our days of female "empowerment." The shift may also reflect a deep Freudian response to the sexual plagues of the time; whatever the motor paralysis of syphilis signified for nineteenth century patients, the chronic sensory fatigue of AIDS may be telling us today. The patients suffer, each in the fashion of the day, each in search of the most radical remedy. Dr. Oliver Wendell Holmes sympathized with the psychosomatic patients of his own day:

> The persons who seek the aid of the physician are very honest and sincere in their wish to get rid of their complaints, and . . . there is nothing they have not done, to recover their health and save their lives. They have submitted to be half-drowned in water, and half-choked with gases, to be buried up to their chins in earth, to be seared with hot irons like galley-slaves, to be crimped with knives, like cod-fish, to have needles thrust into their flesh, and bonfires kindled on their skin, to swallow all sorts of abominations, and to pay for all this, as if to be singed and scalded were costly privilege, as if blisters were a blessing, and leeches were a luxury.[7]

Whatever doctors did to patients in the nineteenth century, patients are capable of doing for themselves in the twenty-first. One's doubts about CFS, ME, or fibromyalgia are reinforced by a recent story from the eccentric mother isle:

> BRITON CURES FATIGUE BY DRILLING HOLE IN OWN HEAD February 22, 2000 LONDON (Reuters)—A British woman says she has cured her chronic fatigue by resorting to do-it-yourself brain surgery and drilling a hole in her own head. Heather Perry, 29, performed the ancient technique of trepanning—cutting away a section of the scalp and drilling into the skull—in her bid to overcome myalgic encephalomyelitis, or ME, which leaves sufferers feeling permanently exhausted.[8]

Perry's bid to rid herself of the inflammation of her brain and spinal chord, by drilling a two-centimeter hole to allow blood to flow more easily around the brain, almost went wrong when she drilled too far and penetrated the dura mater, the membrane protecting her brain tissue. British doctors had refused to help Perry with the ancient procedure, so she flew to an unnamed location in the United States, where she was given medical advice and then did it herself. She said the twenty-minute operation had improved her quality of life.

"I have no regrets. I was prone to occasional bouts of depression and felt something radical needed to be done," said Perry, who performed the operation under local anaesthetic in front of a mirror and a camera crew. (No network was named. I should have thought Oprah or Springer.) "I felt the effects immediately, I can't say they have been particularly dramatic but they are there. I generally feel better and there's definitely more mental clarity. I feel wonderful," she told reporters at her home in Gloucester, western England. Given this success, leeches would indeed have been a luxury for Ms. Perry.

A CANNON-SHOT

Diseases of bodily unrest, like fibromyalgia, certainly cause suffering to those afflicted. But, I'm persuaded that when we finally understand them they'll turn out to have been due to mental errors in what Walter B. Cannon called "the wisdom of the body." Cannon was the twentieth century's recognized leader of physiology in this country, "the Claude Bernard of American science." His major discovery was that a hormone of the adrenal medulla, adrenaline (epinephrine), produced the "fight or flight" response to danger. He was the first to appreciate that several seemingly unrelated changes in the physiology and behavior of an alarmed animal, its breathing pattern, cardiac rhythm, and nervous arousal, were all due to a rush of adrenalin. The notion that simple chemicals link physical reactions to complex emotions had great appeal to the public, and prompted Cannon to produce a stream of articles and reviews for the lay press in which he popularized his theory of homeostasis.

Cannon became even better known for *The Wisdom of the Body*

(1932), a popular book of science, and for *The Way of an Investigator* (1945), a memoir.[9] Not the least of his accomplishments was to forge a link in that chain of Harvard Medical School physician/scientists who were also lucid writers. Their lives spanned almost two centuries, 1809 to 1993; Dr. Oliver Wendell Holmes taught William James who taught Walter B. Cannon who taught Lewis Thomas. Cannon had written his undergraduate thesis at Harvard (1894) on Oliver Wendell Holmes and wrote in the heat of enthusiasm:

> [Holmes's] lecture on homeopathy is conceived and written in a vein of noble scorn and the thought is poured out along the pages with a lucidity, pungency and satire, and cogent understatement that gives the performance the velocity of a cannon-shot.[10]

Cannon-shot, indeed. The young senior could have been describing himself. Torn between a career in medicine and one in philosophy, between the practical and speculative camps of Harvard thought, Cannon went to William James for advice. James dissuaded Cannon from pursuing philosophy: "Don't do it, you will be filling your belly with the East wind."[11] Cannon turned to medicine, and—as it turned out—filled his belly with barium. He made the pioneering discovery that one could visualize the human gastrointestinal tract by means of radio-opaque substances such as barium. It put his reputation on the map. Cannon's later discoveries were closer to the Jamesian temper, a hunt for the general in the "blooming, buzzing confusion" of particulars. By 1914, young Cannon had worked out not only *how* adrenaline affected so many functions of the body, but *why*. He reckoned that the sum of our reactions to epinephrine was to equip us for responding rapidly to danger. His description of the alarm reaction was a physiologist's account of how Darwinian evolution might play itself out in the field:

> The organism which with the aid of increased adrenal secretion can best muster its energies, can best call forth sugar to supply the

laboring muscles, can best lessen fatigue, and can best send blood to the parts essential in the run or the fight for life, is most likely to survive.[12]

Cast into rhyme as the "fight or flight" reaction, the theory was validated in the battalion aid stations and field hospitals of the Great War; the face of fear was drawn by adrenaline.

Cannon broadened his observations on the alarm reaction in 1926, when he formulated one of the more holistic concepts in biology, that of homeostasis, an ingenious Yankee extension of Claude Bernard's theory of the internal milieu. Cannon explained that the Gallic balance of internal fluids, the constancy of our blood, sweat, and tears was maintained by a series of self-regulating pressure/volume circuits under the control of hormones. Those circuits maintained not only the constancy of inner fluids, but permitted the body as a whole to adjust to change:

> The steady states of the fluid matrix of the body are commonly preserved by physiological reactions. . . . Special designations are therefore appropriate: *homeostasis* to designate stability of the organism; *homeostatic conditions* to indicate details of the stability.[13]

It is perhaps fitting that Cannon announced homeostasis in a volume dedicated to the most holistically inclined of William James's colleagues, the Parisian Charles Richet. Richet (1850–1935) was an experimental physiologist who won the Nobel Prize in 1913 for his studies of anaphylaxis—the kind of immediate, immune paroxysm induced by a bee sting, for example, a reaction that can be partially quelled by epinephrine. Richet was also a novelist and playwright, and—like William James—a sometime president of the Society for Psychical Research. He was convinced of telepathy, of telekinesis, and of the reality of ectoplasmic phenomena—a prime example of the Jamesian Will to Believe.[14]

Cannon carried the notions of homeostasis into the social realm. He was a political progressive and devoted much of his energy in the late

1930s to antifascist causes. His rousing speech in support of the Spanish republic, now in the Blodgett collection of Spanish civil war pamphlets at Harvard College Library, remains a vivid document of his passion—and of the period. In accord with these urges, he argued that the concept of homeostasis, properly modified, could be used to explain the human need for material and scientific progress. "The functioning of the human brain has made *social homeostasis* differ markedly from *physiological homeostasis*. ... An upset of constancy necessarily results."[15] In the darkest days of the twentieth century, Cannon was persuaded that the Wisdom of the Body had social significance. He would probably agree that putting a hole in one's own head for "an upset of constancy" shows that a body in despair loses wisdom.

May 8, 2000

Gene Therapy and Sophia Loren

THIS WEEK all the frontline bulletins come from the gene-replacement front. In the first instance, the very technique that the HIV retrovirus uses to enter our cells was turned into a strategy to help sick children.

SUCCESSFUL GENE THERAPY IN HUMANS

In first place is the successful treatment via gene therapy of severe combined immunodeficiency syndrome (SCID) by a group of French clinical investigators.[1] Severe combined immunodeficiency-X1 (SCID-X1) is an X-linked inherited disorder characterized by an early block in lymphocyte differentiation (T and natural killer cells). The disease is caused by mutations in genes of certain molecular messengers of inflammation and immunity, specifically the common subunit of several cytokine receptors. These are receptors without which lymphocytes— and children—fail to thrive. The doctors introduced a harmless retrovirus-derived vector containing the missing gene into the children's lymphoid cells in a dish, and then reinfused the cells into the donors. After a ten-month follow-up period, transgene-expressing T and NK cells were detected in two patients. T, B, and NK cell counts and function, including antigen-specific responses, were comparable to those of age-matched controls.

This form of SCID is only one of a family of immunologic flaws, many of which have no relationship to cytokine receptors. Indeed, the earliest recognized forms of SCID turned out to be due to abnormalities in an obscure branch of intermediate metabolism, the purine salvage pathway. The worst form is due to lack of the enzyme adenosine deaminase.[2]

In the 1970s, Eloise Giblett of Seattle and Rochelle Hirschhorn of NYU, among others, described the clinical syndromes caused by genetic variants in the purine salvage pathway; it was soon found that accumulation of various toxic intermediates such as deoxyadenosine caused defects of lymphocyte growth and function in both the T and B cell series. Cellular and humoral immunity were both depressed; it was *combined* immunodeficiency, and it was certainly severe. Hence the acronym. And it was certainly genetic, thereby differing from acquired forms of immunodeficiency. Indeed, without the definition of SCID, it would have been difficult for investigators in the 1980s to understand almost immediately what they were up against with AIDS, the name of which was appropriately derived from SCID.

There is yet another strange twist to this tale. Among the immediate tributes that greeted this triumph of Parisian medicine, no one mentioned how we came to know so much about SCID, nor the purine salvage pathway, for that matter. The purine salvage pathway was unearthed by biochemists in the '50s and '60s who were studying a far more humble disease, gout.[3] The work of James Wyngaarden, Alexander Gutman, and Gertrude Elion, again, among others, taught us how purines are handled, how their analogues can treat malignancy, and not incidentally, how to put gout in its place. So it's a strange story: from gout to the purine salvage pathway, to SCID, to AIDS, and on to the first real triumph of gene therapy. Success—in gene therapy as in other endeavors—has many parents; failure is an orphan.

GENE THERAPY DEATHS (ANECDOTE AS DATA)

The orphanage is well publicized. As extensively covered in all the media, a young teenager died in Philadelphia last winter after undergoing gene therapy by means of an adenovirus vector. The case forced suspension of similar work at Penn, the convening of special panels, committees, and hearings; it also prompted second thoughts about medical ethics nationwide. Now two more cases have hit the press. Here are the facts, as posted on the FDA web site and reported by the *Washington Post* and the *New York Times* last week.[4,5] A team led by Dr. Jeffrey M. Isner at Tufts/St. Elizabeth's Medical Center in Boston failed to report the death of a patient in gene therapy experiments and might have contributed to the growth of cancer in another patient, whose condition was "reported improperly." In a study funded by a company in which he has an interest, Isner injected the angiogenesis-promoting gene for VEGF (vascular endothelial growth factor) directly into myocardium (heart muscle) of patients with narrowed coronary vessels.[6]

The FDA found several violations of the rules of the experiment in a routine check in March. One patient in the experiment died two months after receiving VEGF, but the Tufts group failed to report the death to the FDA, which is now about to determine if the treatment was responsible. The second patient was a heavy smoker who had a small mass in one lung, the agency said. Because VEGF is angiogenic, and since tumor growth is dependent on angiogenesis, the treatment may have exacerbated the disease. The FDA warning letter on the web notes that St. Elizabeth's doctors spotted the mass when it was less than a centimeter in July 1999, then again in August, when it was two centimeters in diameter, but went ahead with the VEGF therapy on September 21, 1999. Two months later, the mass had grown to five centimeters without evidence either in the patient's records or in the experiment records that either the patient or his doctor was notified— a not uncommon problem in our age of HMOs. This and three other experiments carried on by Dr. Isner were stopped when the agency first began its inspection in February and have not resumed. "Dr. Isner is a founder of the company leading the trials and is a major stock-holder in it," Philip Hilts reported in the *New York Times*.

This case is problematic, to say the least, because tumor growth *could* have been due to the VEGF treatment, or perhaps it was due to the narrow, procedure-oriented medicine we're forced to practice by the HMOs. I'd guess that Isner's two cases, like the one in Philadelphia, are simply the first anecdotes in a long tale that will take time to tell; they are failures in search of a success like the one in Paris.

Secretary of Health and Human Services Donna Shalala had the last word at a press conference in Washington at which she discussed the publicity surrounding gene therapy.[7] Pledging more oversight, but promising no retreat from science, she warned reporters not to generalize from single cases: "In this town, anecdote becomes data." When gene therapy becomes as commonplace as appendectomy, I suppose that the opposite will hold.

THE SOPHIA LOREN EXPERIMENT

Not only viruses can be used as vectors for genes. Liposomes, tiny sacs of test-tube phospholipids that enclose or contain drugs and enzymes, can also be formed to contain DNA. And while it took over fifteen years after they were named before liposomes containing various pharmaceuticals came into clinical use, it took only five years for the houses of Dior and Lancôme to formulate cosmetics in which fragrances, vitamins, and "restoratives" were plonked into the little vesicles.[8]

Indeed, for a while in the '80s, no decent cosmetic house was without its own kind of liposomal preparation to butter the skin. I remember that perpetual face of youth, a younger Sophia Loren, smiling into a camera with a schmeer of liposomes (either Niasome or Capture, I forget which) on her ravishing cheekbones. And this week, gene therapy has come of age in the skin trade. George Cotsarelis and colleagues of Penn's fine Department of Dermatology have put new life into old skin by means of liposome-encapsulated transgenes.[9]

Arguing that topical delivery of transgenes to hair follicles might be used to treat disorders of the skin and hair, they applied liposome-DNA mixtures (to which they gave the cinematic name of "lipoplex") to mouse skin and to human skin planted into nude mice (a xenograft). This resulted in efficient in vivo transfection of hair follicle cells, but only at the onset of a new growing stage of the hair cycle. They manipulated the hair cycle by the kind of rough treatment that movie stars and supermodels undergo almost weekly. "With depilation and retinoic acid," or by means of peeling and vitamin A as they say at Elizabeth Arden, "treatment resulted in nearly 50 percent transfection efficiency, defined as the proportion of transfected, newly growing follicles within the xenograft." The scientists claim that transgenes administered in this fashion are selectively expressed in hair progenitor cells and therefore have the potential to "alter hair follicle phenotype and treat diseases of the hair and skin."[10]

When my bald pate sprouts Neapolitan hair à la Sophia, we'll have data to deal with, not anecdote.

April 3, 2000

Herbal Warnings and the Three Aspirins

HERBS AND SPICES

LAST THURSDAY the Food and Drug Adminstration held a meeting at Gaithersburg to debate whether manufacturers of dietary supplements should be allowed to claim that their "benefits" apply to pregnant women.[1] No conclusions were reached, but it was clear that there is growing concern that these "natural" remedies can carry unnatural side effects. Pharmacists joined in, calling attention to the "unexpected effects of many popular herbal products that are described in the literature."[2]

Now that half the civilized world is gulping food supplements or herbal remedies, with Evian water as a chaser, we might note that these "natural" products can have serious side effects. Recent studies in the *Lancet* warn us that these supplements, which are frequently used as self-prescribed drugs, have not undergone testing by the FDA at the rigorous level of phase three studies required of ethical pharmaceuticals. Dr. Fugh-Berman of George Washington University lists the troubles patients can run into over the counter on their way to filling prescriptions for ethical drugs that actually work.[3]

Herbal medicines may mimic, magnify, or oppose the effect of ethical drugs. Plausible cases of herb-drug interactions include: bleeding when warfarin is combined with ginkgo (*Ginkgo biloba*), garlic (*Allium sativum*), dong quai (*Angelica sinensis*), or danshen (*Salvia miltiorrhiza*); mild serotonin syndrome in patients who mix Saint-John's-wort (*Hypericum perforatum*) with serotonin reuptake inhibitors; decreased efficacy of digoxin, theophylline, cyclosporin, and coumadin when these drugs are combined with Saint-John's-wort; induction of mania in depressed patients who mix antidepressants and Panax ginseng; exacerbation of extrapyramidal effects with neuroleptic drugs and betel nut (*Areca catechu*); increased risk of hypertension when tricyclic antidepressants are combined with yohimbine (*Pausinystalia yohimbe*); potentiation of oral and topical cortisone-like drugs by liquorice (*Glycyrrhiza glabra*); decreased blood concentrations of prednisolone when taken with the Chinese herbal product xaio hu tang (*sho-salko-to*); and decreased concentrations of antiepileptics when combined with the Ayurvedic syrup shankhapushpi. Anthranoid-containing plants (including senna [*Cassia senna*]) interfere with the absorption of several ethical drugs.

More bad news from the herbal front also comes from the *Lancet*. Two studies report that Saint-John's-wort dulls the effectiveness of both indinavir in HIV-infected patients and of cyclosporin in heart transplant patients.[4] In February, the U.S. Food and Drug Administration took note of the studies, cautioning doctors about using Saint-John's-wort with the medications. The FDA said it was working with drug manufacturers to ensure that labeling of the medications be revised "to highlight the potential for drug interactions with Saint-John's-wort." Unfortunately—thanks to powerful congressional lobbies and sentimental fans of alternative therapies—they are powerless to control use of Saint-John's-wort itself. Nor can they regulate the ad lib use of other herbal remedies, poultices, and food supplements such as shark cartilage or tiger teeth.

The FDA does, however, warn patients of toxicities associated with regulated pharmaceuticals, and on February 25, 2000, the FDA warned diabetics to avoid five brands of Chinese herbal products because the herbs illegally contain prescription drugs that could cause severe hypoglycemia. The manufacturers of these "traditional" drugs claim that they contain only natural Chinese herbs, but after a diabetic herb-user had several episodes of hypoglycemia, the California health department discovered the products also contain the prescription diabetes drugs glyburide and phenformin. "People with diabetes should avoid these products and consult their physician if they've been taking them," the FDA warned. The products listed were: Diabetes Hypoglucose Capsules, sold by Chinese Angel Health Products of Santa Monica, CA; Pearl Hypoglycemic Capsules, imported by Sino American Health Products Inc. of Torrance, CA, but also sold by Chinese Angel; Tongyitang Diabetes Angel Pearl Hypoglycemic Capsules and Tongyitang Diabetes Angel Hypoglycemic Capsules, sold by Sino American; Zhen Qi Capsules, sold by Sino American.

The FDA has stopped imports of the products and is investigating how the drugs were added to these preparations. Sino American has agreed to recall the products, which "may be returned to the place of purchase for a refund." No mention is made of where other such products may be returned—or discarded. I remain persuaded that all those unregulated herbal, traditional, naturopathic, homeopathic, etc. remedies are the "materia medica" of our time. Dr. Oliver Wendell Holmes had the last word on this. Holmes told the Mass. Medical Society, 1860:

> Throw out opium, which the Creator himself seems to prescribe, for we often see the scarlet poppy growing in the cornfields, as if it were foreseen that wherever there is hunger to be fed there must also be pain to be soothed; throw out a few specifics which our art did not discover, and is hardly needed to apply; throw out wine, which is a food, and the vapors which produce the miracle of anesthæsia, and I firmly believe that if the whole materia medica, *as now used*, could be sunk to the bottom, it would be all the better for mankind—and all the worse for the fishes.[5]

ASPIRIN: THREE OF THEM

Before doctors hurry to give their elderly patients the newest COX-2 inhibitor (e.g., Enbrel, Vioxx) for hip pain, they may wish to take a look at this week's issue of the *Archives of Internal Medicine*. Doctors Page and Henry of the University of Newcastle in Australia found that use of nonsteroidal anti-inflammatory drugs (NSAIDs) such as aspirin may account for 19 percent of admissions with congestive heart failure.[6] Use of NSAIDs the week before admission doubled the odds of admission for congestive heart failure. But the role of COX-1 and COX-2 in fluid retention is by no means clear, and COX-2 inhibitors can also cause congestive heart failure. On ward rounds, I remind students and house staff that the operative initials in the acronym are A.I.D.(Anti-Inflammatory Drug) and warn them not to give NSAIDs when simple analgesics will do. I often recommend aspirin, the reputation of which is rising, deservedly so.

THE THREE ASPIRINS

I also tell students the history of aspirin—actually of the three aspirins, reviewed in *NSAIDs: Aspirin and Aspirin-like Drugs*. On June 2, 1763, the Royal Society received a communication from Rev. Edward Stone of Chipping Norton in Oxfordshire. Its opening lines are probably unmatched in clinical pharmacology: "Among the many useful discoveries which this age has made, there are very few which better deserve the attention of the public than what I am going to lay before your Lordship. There is a bark of an English tree, which I have found by experience to be a powerful astringent and very efficacious in curing aguish and intermittent disorders."

The tree was the willow (Salix alba), the astringent bark of which contains salicin, the glycoside of salicylic acid. Stone had discovered that salicylates reduced the fever and aches produced by a variety of acute, shiver-provoking illnesses, or agues.

In this year of the millennium, the salicylate most commonly used is acetylsalicylic acid, aspirin. At its lowest dose (80 to 325 mg per day), which we will call aspirin I, acetylsalicylic acid is used to prevent coronary and cerebral thrombosis by virtue of its antiplatelet effect. Over-the-counter doses (650 mg to 3 gr per day), or aspirin II, are used as analgesics and antipyretics (to reduce pain and fever, respectively). Finally, for one hundred years very high doses (> 3 g per day)—aspirin III—have been used to reduce the redness and swelling of joints in rheumatic fever, gout, and rheumatoid arthritis.

ASPIRINS I AND II: FEVER AND PAIN

Aspirin and other salicylates have a variety of other biologic effects, only some of which are related to their use in medicine today. Salicylates can dissolve corns on the toes—a "keratolytic" effect; provoke loss of uric acid from the kidneys—their "uricosuric" property; and kill bacteria in vitro—their antiseptic action. But cell biologists also use aspirin and salicylates to inhibit ion transport across cell membranes, to interfere with the activation of white cells, and to uncouple oxidative phosphorylation by isolated mitochondria. Botanists have found that salicylates serve as host-defense molecules in plants; they also use salicylates to enhance the flowering of *Impatiens*. Indeed salicylates induce the voodoo lily or skunk cabbage to undergo temperature rises of 12 to 16°C in the course of their efflorescence: salicylates therefore not only reduce fever in humans but also produce it in plants. Finally, molecular biologists use salicylates to activate genes that code for heat-shock proteins in the lampbrush chromosomes of *Drosophila*, to influence mitogen-activated protein kinases and to induce programmed cell death (apoptosis) of cancer cells.

"About six years ago," wrote Stone in his letter to the Royal Society, "I accidentally tasted [the willow bark], and was surprised at its extraordinary bitterness; which immediately raised in me a suspicion of its having the properties of the Peruvian bark." Peruvian bark (cinchona) was a venerable remedy for the ague. Stone proceeded to offer a skillful rationale for using willow bark in febrile disorders: the traditional doctrine of signatures—i.e., that "many natural maladies carry their cures along with them, or their remedies lie not far from their cause." Because moist shires, like those drained by the Avon or Isis, abounded in both fevers and willows, Rev. Stone set out to test whether the former might be cured by the latter. Six years of careful clinical observation, and the treatment of fifty patients with willow extracts prepared in water, tea, or beer, culminated in his letter to the Royal Society. The eighteenth century had found a predictable remedy for fever. Hippocrates (fourth century B.C.) had advocated the chewing of willow leaves to relieve the pains of childbirth, and there are references by Pliny (first century) and Galen (second century) to the analgesic property of willow, but it was Stone who put extract of willow bark into our pharmacopoeia as an effective antipyretic agent.

By 1828, at the Pharmacologic Institute of Munich, Büchner isolated a tiny amount of the active glycoside, salicin, in the form of bitter-tasting yellow, needlelike crystals. Two years later, Leroux in Paris improved on the extraction procedure and obtained one ounce of salicin from three pounds of the bark. By 1838, Raffaele Pira of Pisa, writing in the *Comtes Rendu de l'Academie de Science*, described how he obtained a pure substance from salicin by hydrolyzing the glycoside in a CrO_3-mediated oxidation via an aldehyde intermediate. He gave it the name by which we know it today: *"l'acide salicylique,"* or salicylic acid. Willow bark was not alone in providing a rich natural source of salicylates. Meadowsweet (*Spiraea ulmaria*) yielded ample quantities of an ether-soluble oil from which a *Spirsäure* was crystallized in 1835 by the Swiss chemist Karl Jakob Löwig. In 1839 Dumas demonstrated that the *Spirsäure* of Löwig was nothing else than the *acide salicylique* of Pira. Another Gallic pharmacologist, Auguste Andre Thomas Cahours (1843), showed that oil of wintergreen—a traditional remedy for aguish disorders—contained the methyl ester of salicylic acid and prepared *acide salicylique* from it.

As was to be the case in much of nineteenth century chemistry, French and British scientists were slightly ahead of the Germans in the study of natural products, whereas Germans held the edge in synthetic know-how. Forced to compete with the French and British dye industries, which supplied their textile mills with pigments imported from overseas colonies, the Germans responded by inventing cheap aniline dyes, creating in their train such giant enterprises as I. G. Farben. By 1833, the pharmacist E. Merck of Darmstadt had obtained a clean preparation of salicin which was cheaper by half than the impure willow extracts used as antipyretics. But a cheap, pure, acceptable remedy was not available until 1860, when chemist Adolf Kolbe and his students at Marburg succeeded in the first synthesis of salicylic acid and its sodium salt from phenol, CO_2, and sodium. Using industrial variations of the Kolbe synthesis, one of his students, Friedrich von Heyden, established in 1874 the first large factory in Dresden devoted to producing synthetic salicylates. The availability of cheap salicylic acid spread its clinical use far and wide.

ASPIRIN III AND RHEUMATISM

The first successful treatment of acute rheumatism was reported in 1876 by Stricker and Ries in the *Berliner Medizinische Wochenschrifft* and by Maclagan writing in the *Lancet*. Stricker and Ries reported the complete cure of acute "polyarthritis rheumatica" by sodium salicylate at doses of 5 to 6 g per day. Almost simultaneously, Maclagan reported his results with salicylic acid and salicin at similar dosage levels; he paid tribute to the still-prevalent doctrine of signatures, pointing out that cases of acute rheumatism were most abundant in moist areas where the willow grows. Stricker and Ries, and Maclagan, had demonstrated a clinical property of high-dose salicylates which was not tested in the laboratory until the 1930s: They found that salicylates reduce not only fever and pain but also redness and swelling. That anti-inflammatory property was next used to advantage by the Parisian Germain See, who in 1877 introduced salicylates (both acid salicylique and salicin) as effective treatments for gout and "chronic polyarthritis." See had great success among his well-off clientele with using salicylates in acute and chronic gout, so great indeed that the *British Medical Journal*, in an editorial note, called him to task for charging up to £80 to treat a patient with gout by means of 6 to 8 g of sodium salicylate per day, when the price of 1 g of the drug was but five pence. There the matter rested, with high doses of sodium salicylate more or less accepted as a new treatment in many rheumatic diseases, whereas lower doses (1.5 to 2.0 g per day) seemed to relieve aches and pains.

In 1898 a new chapter was written: Felix Hofmann was an aniline-dye chemist at the Friedrich Bayer-Eberfeld division of the I. G. Farben cartel when his father complained to him of gastric irritation from the sodium salicylate he was taking for "rheumatism." Hofmann searched the chemical literature for less acidic derivatives and hit upon acetyl derivatives of sodium salicylate first described by H. von Gilm in 1859 and ten years later by a certain Karl-Johann Kraut. Although von Gilm and Kraut had outlined synthesis of the compound, they had no notion of what its biologic effects might be. Hofmann repeated the synthesis (via acetic anhydride) and tried acetylsalicylic acid first on himself and then on his father: It proved more palatable, less irritating to the stomach, and—he claimed—more effective. Hofmann took the material to his supervisor, Heinrich Dreser, head of Bayer's laboratory of pharmacology, who reported that it performed better in both laboratory and clinic and called the new drug aspirin, the "a" from acetyl and the "spirin" from the German *Spirsaure*.

ALONG CAME SIR JOHN

Unfortunately, until 1971 no useful hypothesis had emerged as to how salicylates exert their various effects. The most important recent contribution to the story of aspirin-like drugs was made by John Vane (now Sir John), then at the Royal College of Surgeons in London, in 1971. Vane had been impressed that many forms of tissue injury are followed by release of prostaglandins (PGs), the oxidation products of arachidonic acid. PGE_1 and PGE_2 had been shown to be associated with acute vasodilation and fever. Vane and his colleagues found that aspirin-like drugs inhibited the biosynthesis of PGE_2 and PGF_{2a} from radiolabeled arachidonic acid in studies in vitro. The enzyme that formed them was called cyclooxygenase, or COX, and aspirin inhibited its action. Just to confuse matters, it turned out there were two forms of this enzyme, COX-1 and 2.[7]

Indeed, current dogma has it that aspirin is anti-inflammatory because it inhibits the two cyclooxygenase enzymes, COX-1 and COX-2, both of which form inflammation-inducing prostaglandins. COX-1 is expressed all the time (constitutively) in many tissues and is important in maintaining the daily activities of our body, what Walter B. Cannon called homeostasis. In contrast, COX-2 is turned on (up-regulated) at sites of need in response to stimuli such as growth factors, immune reactions, and its brothers and sisters in inflammation, the ubiquitous cytokines. Because the products of the COX-1 enzyme regulate stomach acidity and urine output, COX-1 inhibition leads to aspirin's side effects such as stomach irritation and fluid retention. But when Cox-2 is inhibited by newer drugs such as Vioxx or Celebrex, one gets all the benefits of aspirin but with fewer side effects. It's important to note that these COX-2 inhibitors, unlike lifesaving aspirin, have no antiplatelet effect, which is COX-1 dependent.

NOT IN VANE

Recent work from several labs, and not by Vane alone, has complicated this simple formulation. On the one hand, at doses used in rheumatic diseases, aspirin and sodium salicylate block the output of inflammatory substances from white cells by disrupting their intricate web of cellular signalling. On the other hand, aspirin and sodium salicylate retain their anti-inflammatory action in animals which have absolutely no COX-2 whatever because the gene has been deleted. Aspirin inhibits inflammation in COX-2 knockout mice.[8]

Secondly, structural and molecular biologists have come up with the remarkable finding that when aspirin inhibits COX-2, the inhibited enzyme forms a unique set of compounds, called "aspirin-triggered lipoxins" or ATLs. The most recent finding, by a group at Vanderbilt this month, tells us that aspirin inhibits COX activity by acetylating the amino acid serine at the active site (Ser-530) of the enzyme.[9]

Although acetylation of COX-1 and COX-2 occurs on the same residue in the active site, the modified enzymes differ in their product profiles. After aspirin acetylation, COX-1 comes up with nothing, whereas acetylation of COX-2 forces the enzyme to make anti-inflammatory ATLs via an intermediate called 15-(R)-HETE. The work suggests that aspirin is a pro-drug both for salicylate and for the ATLs and it is these metabolites that may be responsible for much of aspirin's effect on cells in a dish.

But, until recently, it was unclear whether these aspirin-induced products were physiologically relevant. The discovery that 15-(R)-HETE can be converted to ATLs (the lipoxins 15 epi-LXA4 and 15 epi-LXB4) in living systems suggests that the beneficial actions of aspirin in animals and humans may indeed be mediated by aspirin-induced generation of the body's own anti-inflammatory products. Serhan's group at Harvard has shown that ATLs have powerful effects in whole human blood and in a variety of inflammatory models, not the least of which is in experimental gingivitis.[10] "Gingivitis!" you might exclaim. But chronic gum disease and its underlying bone erosion looks for all the world like rheumatoid arthritis to a pathologist, and I in turn have always considered rheumatoid arthritis to be gingivitis of the joint. Take two aspirins and call me in the morning on this one.

Notes

INTRODUCTION

1. Allan Nevins and Milton Halsey Thomas, preface to *The Diary of George Templeton Strong* (New York: Macmillan, 1952), p. i.
2. S. Jurvetson, *Red Herring*, no. 99 (June 15–July 1, 2001): 43.
3. Hannah Arendt, "Walter Benjamin, 1892–1940," in *Walter Benjamin "Illuminations,"* ed. H. Arendt (New York: Schocken, 1968), p. 68.
4. Edward M. Kennedy, "Biomedical Sciences in an Expectant Society," in *The Biological Revolution*, ed. G. Weissmann (New York: Plenum, 1979), p. 16.
5. L. Thomas, "closing remarks," *The Biological Revolution*, ed. G. Weissmann (New York: Plenum, 1979), p. 150.
6. C. Connolly, "Deaths from Heart Disease, Cancer, AIDS Declined in 1999," *Washington Post*, June 26, 2001.
7. O. W. Holmes, "For a Meeting of the National Sanitary Society," in *Collected Poems*, vol. 1 (Boston: Riverside Press, 1892), p. 269.
8. Water Supply and Sanitation Collaborative Council, WHO [Water Supply and Sanitation Africa Initiative (WASAI)] http://www.wsscc.org.
9. G. Weissmann, "Puerperal Priority," *Lancet* 349:122–125 (1997).
10. R. W. Emerson, "Carlyle," in *Lectures and Biographical Sketches* (Boston: Riverside Press, 1883), p. 496 (first published in *Scribners Magazine*, May 1881).

OCTOBER 2, 2001

1. R. H. Reid, "Saudi Arabia Cuts Ties with Taliban," AP, September 25, 2001.

2. L. MacNeice, "Among the Turf Stacks in *A Little Treasury of Modern Poetry*, ed. O. Williams (New York: Scribners, 1946), p. 302.

3. W. L. Broad and M. Petersen, "Nation's Civil Defense Could Prove to Be Inadequate Against a Germ or Toxic Attack," *New York Times*, September 23, 2001, p. 12.

4. R. Weiss, "Bioterrorism: An Even More Devastating Threat," *Washington Post*, September 17, 2001, p. A 24.

5. "FBI Imposes New Restrictions on Crop-dusters," cnn.com, September 23, 2001.

6. S. Shafazand, R. Doyle, S. Ruoss, A. Weinacker, T. A. Raffin, "Inhalational Anthrax: Epidemiology, Diagnosis, and Management," *Chest* 116(5): 1369–76 (November 1999).

7. M. Meselson, J. Guillemin, M. Hugh-Jones, et al. "The Sverdlovsk Anthrax Outbreak of 1979," *Science* 266:1202–8 (1994).

8. J. Guillemin, *Anthrax: The Investigation of a Deadly Outbreak* (Berkeley, CA: University of California Press, 1999).

9. L. M. Grinberg, F. A. Abramova, O. V. Yampolskaya, D. H. Walker, and J. H. Smith, "Quantitative Pathology of Inhalational Anthrax I: Quantitative Microscopic Findings," *Mod Pathol* 14(5):482–95 (May 2001).

10. S. F. Little, B. E. Ivins, "Molecular Pathogenesis of Bacillus Anthracis Infection," *Microbes Infect* 1(2):131–39 (February 1999).

11. Y. Singh, K. R. Klimpel, S. Goel, P. K. Swain, S. H. Leppla, "Oligomerization of Anthrax Toxin Protective Antigen and Binding of Lethal Factor During Endocytic Uptake into Mammalian Cells," *Infect Immun* 67(4):1853–59 (April 1999).

12. T. J. Goletz, K. R. Klimpel, N. Arora, S. H. Leppla, J. M. Keith, J. A. Berzofsky, "Targeting HIV Proteins to the Major Histocompatibility Complex Class I Processing Pathway with a Novel Gp120-anthrax Toxin Fusion Protein," *Proc Natl Acad Sci U S A* 94(22):12059–64 (October 28, 1997).

13. T. V. Inglesby, D. A. Henderson, J. G. Bartlett, et al., "Anthrax as a Biological Weapon: Medical and Public Health Management," *JAMA* 281:1735–45 (1999).

14. R. B. Zurier, G. Weissmann, S. Hoffstein, S. Kammermann, and H.-H. Tai, "Mechanisms of Lysosomal Enzyme Release from Human Leukocytes II: Effects of cAMP and cGMP, Autonomic Agonists, and Agents Which Affect Microtubule Function," *J Clin Invest* 53:297–309 (1974).

15. K. R. Klimpel, N. Arora, S. H. Leppla, "Anthrax Toxin Lethal Factor Contains a Zinc Metalloprotease Consensus Sequence Which Is Required for Lethal Toxin Activity," *Mol Microbiol* 13(6):1093–1100 (September 1994). A. Menard, E. Papini, M. Mock, C. Montecucco, "The Cytotoxic Activity of Bacillus Anthracis Lethal Factor Is Inhibited by Leukotriene A4 Hydrolase and Metallopeptidase Inhibitors," *Biochem J* 1,320(Pt 2):687–91 (December 1996).

16. Paul de Kruif, *Microbe Hunters* (New York: Harcourt Brace & Company, 1926) p. 114.

17. W. H. Auden, "New Year Letter (January 1, 1940)," *The Collected Poetry of W. H. Auden* (New York: Random House, 1945) p. 311.

AUGUST 6, 2001

1. S. G. Stolberg, "House Backs Ban on Human Cloning for Any Objective," *New York Times*, July 31, 2001, p. 1.

2. Stolberg, p. 1.

3. Oxford English Dictionary: **embryo** n.1. The offspring of an animal before its birth (or its emergence from the egg): a. of man. In mod. technical language restricted to the fetus *in utero* . . .) [OED http://dictionary.oed.com 1989] • *Le Petit Robert:* **embryon** *n.m. (1361, Gr. embruon) Chez l'homme, Produit de la segmentation de l'oeuf jusqu'à la huiteme semaione du developpement dans l'uterus. [Le Petit Robert Dictionnaire de la Langue Francaise* (Paris, 1989), p. 625] • American Heritage Dictionary: **embryo**. In humans, the prefetal product of conception from implantation through the eighth week of development. [*American Heritage Dictionary*, 2000 http://www.bartleby.com/am]

4. Stolberg, p. 1.

5. M. J. Shamblott et al., "Human Embryonic Germ Cell Derivatives Express a Broad Range of Developmentally Distinct Markers and Proliferate Extensively In Vitro," *Proc Natl Acad Sci USA* 98(1):113–18 (January 2, 2001).

6. http://www.nih.gov/news/stemcell/fullrptstem.pdf.

7. V. de Dios, J. Correa, E. Sgreccio, "Dichiarazione sulla produzione e sull' uso scientifico e terapeutico delle cellule staminali embrionali umane," *Pontificia Accademia per la Vita* (August 2, 2000).

8. A. Stanley, "Bush Hears Pope Condemn Research in Human Embryos," *New York Times*, July 23, 2001, p. 1.

9. Stolberg, p. 1.

10. J. Loeb, "On Artificial Parthenogenesis in Sea Urchins," *Science* 11:612–14 (1900).

11. Y. Delage and M. Goldsmith, *La parthénogénèse naturelle et expérimentale* (Paris: Flammarion, 1913), p. 302.

12. J. Loeb, *Comparative Physiology of the Brain and Comparative Psychology* (New York: G. P. Putnam's Sons, 1900), p. 287.

13. "Reproduction of Humans," *Boston Evening Transcript*, October 2, 1900.

14. G. Pincus, *The Eggs of Mammals* (New York: Macmillan, 1936).

15. P. J. Pauly, *Controlling Life Jaques Loeb and the Engineering Ideal in Biology* (Oxford and New York: Oxford University Press, 1987), p. 191.

16. Pincus, p. 111.

17. Ibid.

18. Pincus, p. 112.

19. "Brave New World," *New York Times*, March 28, 1936, quoted in Pauly, p. 180.

20. J. D. Ratliff, "No Father to Guide Them," *Collier's*, March 20, 1937, p. 137, quoted in B. Asbell, *The Pill* (New York: Random House, 1997), p. 120.

21. K. Nakagawa et al., "A Combination of Calcium Ionophore and Puromycin Effectively Produces Human Parthenogenones with One Haploid Pronucleus," *Zygote* (1):83–8 (February 2001).

22. E. M. Ongeri, C. L. Bormann, R. E. Butler, D. Melican, W. G. Gavin, Y. Echelard, R. L. Krisher, and E. Behboodi, "Development of Goat Embryos after in Vitro Fertilization and Parthenogenetic Activation by Different Methods," *Theriogenology* 55(9):1933–51 (June 1, 2001).

23. A. R. Weeks, F. Marec, and J. A. J. Breeuwer, "A Mite Species That Consists Entirely of Haploid Females," *Science* 292:2479–82 (June 29, 2001).

24. C. Zimmer, "Wolbachia: A Tale of Sex and Survival," *Science* 292(5519): 1093–5 (May 11, 2001).

25. R. Stouthamer, J. A. Breeuwer, G. D. Hurst, "*Wolbachia pipientis*: Microbial Manipulator of Arthropod Reproduction," *Annu Rev Microbiol* 53:71–102 (1999).

26. M. E. Huigens et al., "Infectious Parthenogenesis," *Nature* 405(6783):178–89 (May 11, 2000).

27. A. Hoerauf et al., "Endosymbiotic Bacteria in Worms as Targets for a Novel Chemotherapy in Filariasis," *Lancet* 355(9211):1242–43 (April 8, 2000).

28. N. W. Brattig, D. W. Buttner, and A. Hoerauf, "Neutrophil Accumulation

Around Onchocerca Worms and Chemotaxis of Neutrophils Are Dependent on Wolbachia Endobacteria," *Microbes Infect* 3(6):439–46 (May 2001).

JULY 9, 2001

1. L. Altman, "Artificial-Device Recipient Said to Make Excellent Progress," *New York Times*, July 5, 2001, p. 1.

2. E. Schmitt, "Cheney, Back at Work, Says He's Feeling 'Pretty Good,' " *New York Times*, July 3, 2001, p. 16.

3. J. Traub, "Sierra Leone: The Worst Place on Earth," *NY Review of Books* 47(11): 61–66 (June 29, 2000).

4. "RUF Frees Another 131 Child Soldiers," Abidjan, June 25 (IRIN) http://www.reliefweb.int/IRIN/wa/countrystories/sierraleone/20010625. phtml.

5. B. Ramster, "Aventis Sign Deal with WHO to Supply Sleeping Sickness Drug," *Trends Parasitol* 17(7):312 (July 2001).

6. http://www.msf.org/countries/page.cfm?articleid=3A09D5DA-E42A-11D4-B2010060084A6370.

7. D. Bruce, "Further Report on Sleeping Sickness in Uganda," *Rep Sleeping Sickness Comm R Soc* 4:3–6 (1903), quoted in G. Hide, "History of Sleeping Sickness in East Africa," *Clin Microbiol Rev* 12(1):112–25 (January 1999).

8. http://www.nobel.se/medicine/laureates/1907/press.html.

9. http://www.msf.org/advocacy/accessmed/diseases/sleeping/reports/1999/11/stats.htm.

10. M. Shimamura, K. M. Hager, and S. L. Hajduk, "The Lysosomal Targeting and Intracellular Metabolism of Trypanosome Lytic Factor by Trypanosoma Brucei Brucei," *Mol Biochem Parasitol* 115(2):227–37 (July 2001).

11. J. W. W. Stephens and H. B. Fantham, "On the Peculiar Morphology of a Trypanosome from a Case of Sleeping Sickness and the Possibility of Its Being a New Species (T. rhodesiense)," *Ann Trop Med Parasitol* 4:343–50 (1910).

12. World Health Organization, "New Lease of Life for Resurrection Drug," http://www.who.int/tdr/publications/tdrnews/news64/eflornithine.html.

13. K. Gull, "The Biology of Kinetoplastid Parasites: Insights and Challenges from Genomics and Postgenomics," *Int J Parasitol*, 31(5–6):443–52 (May 2001).

14. H. Denise and M. P. Barrett. "Uptake and Mode of Action of Drugs Used Against Sleeping Sickness: Sleeping Sickness Is Resurgent in Africa," *Biochem Pharmacol* 61(1):1–5 (January 1, 2001).

15. G. A. Cross, "Antigenic Variation in Trypanosomes: Secrets Surface Slowly," *Bioessays* 18(4):283–91 (April 1996).

16. H. Ngo et al., "Double-Stranded RNA Induces mRNA Degradation in Try-panosoma Brucei," *Proc Natl Acad Sci USA* 95(25):14687–89 (December 8, 1998).

17. S. Boseley, "Drug Firm Wakes Up to Sleeping Sickness," *Guardian* (UK), May 7, 2001, p. 3.

18. C. J. Bacchi, "Content, Synthesis, and Function of Polyamines in Trypanoso-matids: Relationship to Chemotherapy," *J Protozool* 28(1):20–27 (February 1981).

19. J. Pepin et al., "Short-course Eflornithine in Gambian Trypanosomiasis: A Multicentre Randomized Controlled Trial," *Bull World Health Organ* 78(11): 1284–95 (2000).

20. P. I. Hynd and M. J. Nancarrow, "Inhibition of Polyamine Synthesis Alters Hair Follicle Function and Fiber Composition," *J Invest Dermatol* 106(2): 249–53 (February 1996). J. Shapiro and H. Lui Vaniqa—Eflornithine 13.9% cream. 1: *Skin Therapy Lett* 6(7):1–5 (April 2001).

21. D. McNeil Jr., "Profits on Cosmetic Save a Cure for Sleeping Sickness," *New York Times*, February 9, 2001, p. 4.

22. PR Newswire, "World Health Organization and Aventis Announce a Major Initiative to Step Up Efforts against Sleeping Sickness," May 3, 2001.

23. Traub, pp. 61–66.

24. F. Irmay, B. Merzouga, and D. Vettorel, "The Krukenberg Procedure: A Sur-gical Option for the Treatment of Double Hand Amputees in Sierra Leone," *Lancet* 356(9235):1072–75 (September 2000).

25. "Plastic Surgeons to Remove Children's Scars," Abidjan, June 20, 2001, (IRIN) http://www.reliefweb.int/IRIN/wa/countrystories/sierraleone/ 20010620.phtml.

26. World Health Organization, "Water Supply and Sanitation Africa Initiative (WASAI)," http://www.wsscc.org.

27. O. W. Holmes, *A Mortal Antipathy*, vol. 7, *Collected Works* (Boston: River-side Press, 1892), p. 171.

28. O. W. Holmes, "For a Meeting of the National Sanitary Society," in *Collected Poems*, vol. 1 (Boston: Riverside Press, 1892), p. 269.

JUNE 12, 2001

1. "George Bush, le président qui remet de l'arsenic dans l'eau potable," *Le Monde*, April 21, 2001.

2. Jack Perlez and Frank Bruni, "Bush Trip Aimed at Winning over Europeans," *New York Times*, June 9, 2001, p. 1.

3. Wendy Williams, "Survival of the Short List: Bush Closes the Wallet on the Endangered Species Act," http://www.prospect.org/webfeatures/2001/04/williams-we-04-20.html.

4. "The Arsenic Ruckus," *Wall Street Journal*, April 25, 2001, Review & Outlook. Also: "The Arsenic Ruckus," http://www.junkscience.com/apr01/wsj=_arsenic.htm.

5. Patrick Connole, "Group to Sue Bush over Suspension of Arsenic Rule," Reuters, Washington, May 22, 2001.

6. http://books.nap.edu/books/0309063337/html/1.html.

7. http://search.npr.org/cf/cmn/cmnpd01fm.cfm?PrgDate=06/06/2001&Prg ID=3.

8. K. Yamanaka et al., "Oral Administration of Dimethylarsinic Acid, a Main Metabolite of Inorganic Arsenic, in Mice Promotes Skin Tumorigenesis Initiated by Dimethylbenz(a)anthracene with or Without Ultraviolet B as a Promoter," *Biol Pharm Bull* 24(5):510–14 (May 2001).

9. G. Stohrer, "Arsenic Levels Can Be Standard or Safe," *Science* 292(5520):1299–300 (May 2001).

10. "Napoléon a-t-il été empoisonné à l'arsenic, puis achevé au sirop d'orgeat?" *Le Monde*, June 2, 2001.

11. Sten Forshufvud, *Who Killed Napoleon?* (London: Hutchinson, 1961).

12. Ben Weider, *The Murder of Napoleon* (New York: Congdon & Lattes, 1982).

13. B. Weider and J. H. Fournier, "Activation Analyses of Authenticated Hairs of Napoleon Bonaparte Confirm Arsenic Poisoning," *Am J Forensic Med Pathol* 20(4):378–82 (December 1999).

14. Jon Henley, "How Perfidious Albion May Have Poisoned Napoleon," *Guardian* (Manchester), May 5, 2000.

15. D. E. H. Jones and K. W. L. Ledingham, "Arsenic in Napoleon's Wallpaper," *Nature* 299:626–67 (October 14, 1982).

16. http://www.napoleon.org/us/us_cd/bib/articles/textes/arsenic.html.

17. Henley.

18. N. Bonaparte, in *Citations Francaise*, ed. Pierre Oster (Paris: Le Robert, 1992), p. 428.

19. S. Waxman and K. C. Anderson, "History of the Development of Arsenic Derivatives in Cancer Therapy," *Oncologist* 6, suppl. 2:3–10 (2001).

20. http://www.fda.gov/cder/approval/index.htm.

21. PR Newswire, "Studies Show TRISENOX™ May Fill an Unmet Need for Relapsed or Refractory Multiple Myeloma Patients," May 4, 2001.

22. B. J. Druker et al., "Activity of a Specific Inhibitor of the BCR-ABL Tyrosine Kinase in the Blast Crisis of Chronic Myeloid Leukemiaand Acute Lymphoblastic Leukemia with the Philadelphia Chromosome," *N Engl J Med* 344(14):1038–42 (April 5, 2001).

23. M. Porosnicu et al., "Co-treatment with As2O3 Enhances Selective Cytotoxic Effects of STI-571 Against BRC-ABL-positive Acute Leukemia Cells," *Leukemia* 15(5):772–78 (May 2001).

24. O. Sordet et al., "Mitochondria-targeting Drugs Arsenic Trioxide and Lonidamine Bypass the Resistance of TPA-differentiated Leukemic Cells to Apoptosis," *Blood* 97(12): 3931–40 (June 15, 1997).

25. H. D. Sun et al., "Ai-Lin I Treated 32 Cases of Acute Promyelocytic Leukemia," *Chin J Integrat Chin & West Med* 12:170 (1992).

26. P. Zhang et al., "Arsenic Trioxide Treated 72 Cases of Acute Promyelocytic Leukemia," *Chin J Hematol* 2:58 (1996).

27. S. L. Soignet et al., "Complete Remission After Treatment of Acute Promyelocytic Leukemia with Arsenic Trioxide," *N Engl J Med* 339(19):1341–48 (November 1998).

28. P. Ehrlich, "Chemotherapy" (from Proc 17th Int. Congress Med. 1913), in *The Collected Papers of Paul Ehrlich*, vol. 3, ed. F. Himmelweit (London: Pergamon Press, 1960) pp. 505–18.

MAY 16, 2001

1. Reuters, "U.S. Approves Novartis Cancer Drug Gleevec," May 10, 2001, Washington.

2. L. Weissmann, personal communication.

3. B. J. Druker et al., "Efficacy and Safety of a Specific Inhibitor of the BCR-ABL Tyrosine Kinase in Chronic Myeloid Leukemia," *N Engl J Med* 344(14):1031–37 (April 2001).

4. http://www.nobel.se/medicine/laureates/1908/ehrlich-bio.html.

5. Ibid.

6. S. A. Okie, "New Drug Shows Promise in Battling Form of Leukemia: Pill Is

Model of How to Kill Cancer without Poisoning Cells," *Washington Post*, April 5, 2001, A2.

7. B. J. Druker et al., "Activity of a Specific Inhibitor of the BCR-ABL Tyrosine Kinase in the Blast Crisis of Chronic Myeloid Leukemia and Acute Lymphoblastic Leukemia with the Philadelphia Chromosome," *N Engl J Med* 344(14):1038–42 (April 2001).

8. Okie, p. A2.

9. P. C. Nowell and D. A. Hungerford, "Minute Chromosome in Human Chronic Granulocytic Leukemia," *Science* 132 (3438):1497–99 (1960).

10. J. D. Rowley, "A New Consistent Chromosomal Abnormality in Chronic Myelogenous Leukaemia Identified by Quinacrine Fluorescence and Giemsa Staining," *Nature* 243(5405):290–93 (June 1973).

11. Y. Ben-Neriah et al., "The Chronic Myelogenous Leukemia-Specific P210 Protein Is the Product of the BCR/ABL Hybrid Gene," *Science* 233(4760):212–14 (July 1986). T. G. Lugo et al., "Tyrosine Kinase Activity and Transformation Potency of BCR-ABL Oncogene Products," *Science* 247:1079–82 (1990).

12. T. Meyer et al., "A Derivative of Staurosporine (CGP-41-251) Shows Selectivity for Protein-Kinase C Inhibition and In Vitro Anti-proliferative As Well As In Vivo Anti-tumor Activity," *Int J of Cancer* 43(5): 851–56 (May 1989).

13. A. Barnard, "Five Honored for Work on Leukemia Drug Is Promising for Disease That Hits 5,000 a Year," *Boston Globe*, May 2, 2001, p. A8.

14. S. Crotty, *Ahead of the Curve: David Baltimore's Life in Science* (San Francisco: Univ. of California Press, 2001).

15. P. A. Baeuerle and D. Baltimore, "Activation of DNA-Binding Activity in an Apparently Cytoplasmic Precursor of the Nf-Kappa-B Transcription Factor," *Cell* 53(2):211–17 (1988).

16. L. Thomas, "Endotoxin," in *The Youngest Science* (New York: Viking Press, 1983), p. 151.

17. Thomas L. Papain, "Vitamin A, Lysosomes and Endotoxin: An Essay on Useful Irrelevancies in the Study of Tissue Damage," *Arch Int Med* 110: 782–86 (1962).

18. R. A. Good and L. Thomas, "Studies on the Generalized Shwartzman Reaction IV: Prevention of the Local and Generalized Shwartzman Reaction with Heparin," *J Exp Med* 97:877–88 (1953).

19. G. Weissmann and L. Thomas, "Studies on Lysosomes I: The Effects of Endo-

toxin, Endotoxin Tolerance, and Cortisone on the Release of Acid Hydrolases from a Granular Fraction of Rabbit Liver," *J Exp Med*, 116:433–50 (1962).

20. Papain, pp. 782–86.

21. N. Auphan et al., "Immunosuppression by Glucocorticoids: Inhibition of NF-kB Activity Through Induction of IkB Synthesis," *Science* 270, 286–90 (October 13, 1995).

22. S. C. Kimmel et al., "A mechanism for the Antiinflammatory Effects of Corticosteroids: The Glucocorticoid Receptor Regulates Leukocyte Adhesion to Endothelial Cells and Expression of ELAM-1 and ICAM-1," *Proc Natl Acad Sci USA* 89:9991–95 (November 1, 1992).

23. http://www.booknotes.org/transcripts/10204.htm.

24. B. Ehrenreich, "Science, Lies and the Ultimate Truth," *Time* 137, May 20, 1991, p. 66.

25. Lynn Phillips, "Can This Science Be Saved?" review of recent books by Carl Sagan, Stephen Jay Gould, Gerald Weissmann, *Nation* 262, May 20, 1996, p. 25.

26. D. J. Kevles, *The Baltimore Case: A Trial of Politics, Science, and Character* (New York: W. W. Norton, 1998).

27. M. Marquardt, *Paul Ehrlich* (New York: Schuman, 1951), pp. 232–38.

28. http://nobel.sdsc.edu/medicine/laureates/1908/press.html.

29. J. G. Leyden, "From Nobel Prize to Courthouse Battle: Paul Ehrlich's 'Wonder Drug' for Syphilis Won Him Acclaim but Also Led Critics to Hound Him," *Washington Post*, July 27, 1999.

30. Marquardt, pp. 232–38.

31. Deutsches Historisches Museum, Berlin, http://www.dhm.de/lemo/html/biografien/EhrlichPaul.

32. Marquardt, pp. 232–38.

33. Leyden.

APRIL 3, 2001

1. R. F. Massung et al., "Epidemic Typhus Meningitis in the Southwestern United States," *Clin Infect Dis* 32(6):979–82 (March 2001).

2. S. F. Whiteford, J. P. Taylor, and J. S. Dumler, "Clinical, Laboratory, and Epidemiologic Features of Murine Typhus in 97 Texas Children," *Arch Pediatr Adolesc Med* 155(3):396–400 (March 2001).

3. N. H. Cho et al., "Expression of Chemokine Genes in Human Dermal

Microvascular Endothelial Cell Lines Infected with Orientiatsutsuga-mushi," *Infect Immun* 69(3):1265–72 (March 2001). D. H. Walker, J. P. Olano, and H. M. Feng, "Critical Role of Cytotoxic T Lymphocytes in Immune Clearance of Rickettsial Infection," *Infect Immun* 69(3):1841–46 (March 2001).

4. W. C. Nierman et al., "Complete Genome Sequence of Caulobacter Crescentus," *Proc Natl Acad Sci USA* 98(7):4136–41 (March 2001).

5. H. Zinsser, *Rats, Lice and History* (Boston: Little, Brown, 1935).

6. M. Muller, *Anne Frank: The Biography*, trans. Robert Kimber and Rita Kimber (New York: Henry Holt, 1998). http://www.nizkor.org/features/qar/qar55.html. http://www.nizkor.org/ftp.cgi/camps/bergen-belsen/ftp.py?camps/bergen-belsen//images/belsen01.jpg. http://www.nizkor.org/ftp.cgi/camps/bergen-belsen/ftp.py?camps/bergen-belsen//images/belsen05.jpg.

7. Whiteford et al., 396–400.

8. Ibid.

9. Ibid.

10. D. Raoult and V. Roux, "The Body Louse as a Vector of Reemerging Human Diseases," *Clin Infect Dis* 29(4):888–911 (October 1999).

11. E. Thofern, "The Success of Hygiene in the Last 40 Years," *Zentralblatt fur Bakteriologie, Mikrobiologie und Hygiene [B]* 187(4–6):271–94 (April 1989).

12. D. Raoult et al., "Jail Fever (Epidemic Typhus) Outbreak in Burundi," *Emerg Infect Dis* 3(3):357–60 (July–September 1997).

13. Raoult and Roux, pp. 888–911.

14. Cho et al., 1265–72. T. S. Walker, "Endothelial Prostaglandin Secretion: Effects of Typhus Rickettsiae," *J Infect Dis* 162(5):1136–44 (November 1990). F. M. Rollwagen et al., "Mechanisms of Immunity to Infection with Typhus Rickettsiae: Infected Fibroblasts Bear Rickettsial Antigens on Their Surfaces," *Infect Immun* 50(3):911–16 (December 1985).

15. Walker et al., pp. 1841–46.

16. L. Margulis, *Origin of Eukaryotic Cells* (New Haven: Yale Univ. Press, 1970).

17. S. G. Andersson et al., "The Genome Sequence of Rickettsia Prowazekii and the Origin of Mitochondria," *Nature* 396(6707):133–40 (November 1998).

18. Nierman et al., pp. 4136–41.

19. L. Shapiro and R. Losick, "Dynamic Spatial Regulation in the Bacterial Cell," *Cell* 100(1):89–98 (January 2000).

20. R. Taylor and A. Rieger, "Medicine as a Social Science: Rudolph Virchow on the Typhus Epidemic of Upper Silesia," *Int J Health Serv* 15(4): 547–59:(1985).

21. W. Lindwer, *The Last Seven Months of Anne Frank* (New York: Anchor, 1992, reprint).

22. M. Schlu, "Anne Frank: Projekt zum Nationalsozialismus," March 11, 2001 http://www.martinschlu.de/Nationalsozialismus/annefrank/15ende.htm.

23. Ibid.

24. http://www.nizkor.org/ftp.cgi/camps/bergen-belsen/ftp.py?camps/bergen-belsen//images/belsen01.jpg.

25. http://motlc.wiesenthal.com.

26. "The Big Extract: On Arriving at the Concentration Camp" (extracted from Index on Censorship), *Guardian* (Manchester), June 13, 1998, p. 12.

27. Ibid.

28. Ibid.

MARCH 21, 2001

1. Melody Petersen and Donald G. McNeil Jr., "Maker Yielding Patent in Africa for AIDS Drug," *New York Times*, March 15, 2001, p. 1.

2. S. A. Bozzette et al., "Expenditures for the Care of HIV-infected Patients in the Era of Highly Active Antiretroviral Therapy," *N Engl J Med* 344(11):817–23 (March 2001). K. A. Freedberg et al., "The Cost Effectiveness of Combination Antiretroviral Therapy for HIV Disease," *N Engl J Med* 344(11):824–31 (March 2001).

3. J. Medawar and D. Pyke, *Hitler's Gift: Scientists Who Fled Nazi Germany* (London: Richard Cohen Books, 2000).

4. E. Chargaff, "Chemical Specificity of Nucleic Acids and Mechanism of Their Enzymic Degradation," *Experientia* 6:201–9 (1950).

5. J. Henkel, "Attacking AIDS with a 'Cocktail' Therapy," *FDA Consumer*, July–August 1999.

6. Bozzette et al., pp. 817–23.

7. Julian Borger, "Campus Revolt Challenges Yale over $40m AIDS Drug: University Claims Its Hands Are Tied by Deal with Drug Firm," *Guardian* (Manchester); March 13, 2001, p. 3.

8. W. H. Prusoff, "Substitution of DNA with Base Analogs," in *The Interaction of Drugs and Subcellular Components of Animal Cells*, ed. P. N. Campbell (London: J & A Churchill, 1968), pp. 45–70.

9. Borger, p. 3.

10. Petersen and McNeil, p. 1.

11. AP, "Africa to Get AIDS Drugs Below Cost," March 15, 2001, New York.

12. Ibid.

13. Prusoff, pp. 45–70.

14. Robert Weissman, "AIDS and Developing Countries: Democratizing Access to Essential Medicines," http://www.foreignpolicy-infocus.org/index.html07/01/99.

15. Bozzette et al., pp. 817–23.

16. Freedberg et al., pp. 824–31.

17. L. Thomas, *Lives of a Cell* (New York: Viking, 1970), p. 36.

18. G. Weissmann, "Effects on Lysosomes of Drugs Useful in Connective Tissue Diseases," in *The Interaction of Drugs and Subcellular Components of Animal Cells*," ed. P. N. Campbell (London: J & A Churchill, 1968), pp. 203–17.

19. E. Chargaff, *Essays on Nucleic Acids* (Amsterdam: Elsevier, 1963).

20. R. Hirschhorn et al., "Template Activity of Nuclei from Stimulated Lymphocytes," *Nature* 222:1247–50 (1969).

21. J. Watson, *The Double Helix* (New York: Scribner's, 1968), p. 128.

22. Chargaff, *Chemical Specificity*, pp. 201–9.

23. O. T. Avery, C. M. Macleod, and M. McCarty, "Studies on the Chemical Nature of the Substance Inducing Transformation of Pneumococcal Types: Induction of Transformation by a Desoxyribonucleic Acid Fraction Isolated from Pneumococcus Type III," *J Exp Med* 79: 137–58 (1944).

24. T.-S. Lin, R. F. Schinazi, and W. H. Prusoff, "Potent and Selective In-Vitro Activity of 3'-Deoxythymidin-2'-Ene(3'-Deoxy-2',3'-Didehydrothymidine) Against Human Immunodeficiency Virus," *Biochem Pharmacol* 36(17): 2713–18 (1987).

25. E. M. August et al., "Initial Studies on the Cellular Pharmacology of 3'-Deoxythymidin-2'-ene (d4T): A Potent and Selective Inhibitor of Human Immunodeficiency Virus," *Biochem Pharmacol* 37(23):4419–22 (December 1988).

MARCH 7, 2001

1. "Afghans Reject Appeals to Spare Statues," *New York Times*, March 6, 2001, p. 6.

2. Reuters, "Europe Slaughter Begins on Foot-and-Mouth Worries," February 26, 2001, London. Mike Collett-White, "UK Extends Ban, Hopes to Contain Foot-and-Mouth," Reuters, February 27, 2001, London.

3. Paul Meller, "Belgium Farmers Protest," *New York Times*, February 27, 2001.

4. Alan Cowell, "Britain Moves to Curtail Spread of Foot-and-Mouth Disease," *New York Times*, February 24, 2001, p. 3.

5. F. Brown, "Foot-and-Mouth Disease—One of the Remaining Great Plagues," *Proc R Soc Lond B Biol Sci* 229(1256):215–26 (December 1986).

6. K. Bauer, "Foot-and-Mouth Disease as Zoonosis," *Arch Virol*, suppl. 13:95–97 (1997).

7. Chris Fontaine, "Irish Nix St. Patrick's Day Parade," AP, March 2, 2001, London. http://www.guardian.co.uk/footandmouth/story/0,7369,443686,00.html.

8. Fontaine.

9. http://www.guardian.co.uk/footandmouth/story/0,7369.443686,00.html.

10. A. I. Donaldson and T. R. Doel, "Foot-and-Mouth Disease: The Risk for Great Britain after 1992," *Vet Rec* 131(6):114–20 (August 1992).

11. Bauer, pp. 95–97.

12. S. Neff, P. W. Mason, and B. Baxt, "High-Efficiency Utilization of the Bovine Integrin Alpha(v)Beta(3) as a Receptor for Foot-and-Mouth Disease Virus Is Dependent on the Bovine Beta(3) Subunit," *J Virol* 74(16):7298–306 (August 2000).

13. J. Seipelt et al., "The Structures of Picornaviral Proteinases," *Virus Res* 62(2):159–68 (August 1999).

14. W. Glaser and T. Skern, "Caspases Are Not Involved in the Cleavage of Translation Initiation Factor eIF4GI during Picornavirus Infection," *J Gen Virol* 81, pt. 7:1703–7 (July 2000).

15. David Evans, "EU Vets Extend UK Foot-and-Mouth Export Ban," Reuters, February 27, 2001, Brussels.

16. Donaldson and Doel, pp. 114.–20.

17. John Vidal, "Foot and Mouth Crisis: Global Disease on the Rise: Causes: Finger Pointed at Illegal Trade," *Guardian* (Manchester), February 23, 2001, p. 4.

18. Ibid.

19. http://www.bild.de/service/archiv/2001/mar/01/aktuell/seuche/seuche.html.

20. Rudyard Kipling, "The White Man's Burden," *McClure's Magazine* 12 (February 1899).

21. Joseph Bell, "The Adventures of Sherlock Holmes," *The Bookman*, December 1892, pp. 50–54.

FEBRUARY 21, 2001

1. D. Baltimore, "Our Genome Unveiled," *Nature* 409:814–16 (2001).

2. S. Brenner, "Hunting the Metaphor," *Science* 291:1265–66 (2001).

3. J. C. Venter et al., "The Sequence of the Human Genome," *Science* 291: 1304–51 (2001). http://www.sciencemag.org/genome2001. http://publica tion.celera.com. Sequence Interpretation: http://www.ncbi.nlm.nih.gov:80/ blast/Blast.cgi.

4. E. Lander et al., "The Genome International Sequencing Consortium: Initial Sequencing and Analysis of the Human Genome," *Nature* 409: 860–921 (2001). http://www.nature.com/genome. http://www.nature.com/genomics. http://genome.cse.ucsc.edu.

5. "On Human Nature," *The Economist*, February 17, 2001, p. 79.

6. Tim Radford, "Third World Rush for Human Genome," *Guardian* (London), February 13, 2001, p. 10.

7. "Gene Project Delay Seen—Lack of Funds/Director Says U.S. Won't Be Able to Decipher Entire Human Code by Year 2005 as Hoped," *San Francisco Chronicle*, October 9, 1993.

8. Aaron Zitner, " 'Whole-Genome Shotgun' Missed Its Mark; Research: Celera Head J. Craig Venter Had Big Plans for His Mapping Technique. But It Didn't Work," *Los Angeles Times*, February 11, 2001, p. 44.

9. N. Wade, "Scientists Announce First Interpretations of Human Genetic Sequence," *New York Times*, February 12, 2001, p. 1.

10. Antonio Regalado, "Whitehead's Lander Kept Public Sector in Gene Race," *Wall Street Journal*, February 12, 2001, p. 12.

11. Tim Radford, "Genome Project: Articles of Faith Lie at Heart of Bitter Feud: Rivals Dispute over Method Dog Research Project," *Guardian* (Manchester), February 12, 2001, p. 1.

12. B. R. Jasny and D. Kennedy, "The Human Genome," *Science* 291:1153 (2001).

13. http://www.sciencemag.org/genome2001. http://publication.celera.com. Sequence Interpretation: http://www.ncbi.nlm.nih.gov:80/blast/Blast.cgi. http://www.nature.com/genome. http://www.nature.com/genomics. http:// genome.cse.ucsc.edu.

14. Baltimore, pp. 814–16.

15. S. J. Gould, "Humbled by the Genomes's Mysteries," *New York Times*, February 19, 2001, p. 15.

16. Brenner, pp. 1265–66.

17. C. Darwin, *On the Origin of Species* (1859; reprint, New York: Heritage Press, 1963), p. 445.

18. http://tuna.uchicago.edu.

19. Ibid.

20. A. M. Wilson, *Diderot* (New York: Oxford University Press, 1972), pp. 358–69.

21. Ibid, p. 362.

22. R. Hooke, *Preface to Micrographia* (1665; reprint, New York: Dover, 1962), p. xii.

FEBRUARY 7, 2001

1. C. W. Dugger and H. Kumar, "Quake's Victims: Rich and Homeless, Poor and Homeless," *New York Times*, February 2, 2001, p. 3.

2. N. MacFarquhar, "Qaddafi Rants Against the U.S. in a Welcoming After Bomb Trial," *New York Times*, February 2, 2001, p. 1.

3. J. P. Dworkin et al., "Self-Assembling Amphiphilic Molecules: Synthesis in Simulated Interstellar/Precometary Ices," *Proc Natl Acad Sci USA* 98: 815–19 (January 2001).

4. http://www.nobel.se/medicine/laureates/1959.

5. A. D. Bangham, M. M. Standish, and J. C. Watkins, "Diffusion of Univalent Ions Across the Lamellae of Swollen Phospholipids," *J Mol Biol* 13:238 (1965).

6. R. Hooke, *Preface to Micrographia* (1665; reprint, New York: Dover, 1962), p. xi.

7. David Perlman, "Chemicals Almost Come Alive; NASA Creates Cell-like Membranes by Irradiating Inert Compounds," *San Francisco Chronicle*, January 30, 2001, p. A2.

8. Kathy Sawyer, "In Space, Clues to the Seeds of Life: Chemical 'Membranes' Could Revise Thinking on Origins," *Washington Post*, January 30, 2001, p. A1.

9. Dworkin et al., pp. 815–19.

10. G. Weissmann, "Introduction to Cell Membranes," in *Cell Membranes: Biochemistry, Cell Biology and Pathology*, ed. G. Weissmann and R. Claiborne (New York: HP Press, 1975).

11. http://www.nigms.nih.gov/news/science_ed/surface.html.

12. Sawyer, p. A1.

13. D. W. Deamer, "A Visit to England," in *The Liposome Letters*, ed. A. D. Bangham (New York and London: Academic Press, 1983), p. 1.

14. D. W. Deamer, "The First Living Systems: A Bioenergetic Perspective," *Microbiol Mol Biol Rev* 61:239–61 (June 1997).

15. Lucretius, *On the Nature of Things*, trans. F. O. Copley (circa 60 BCE; New York: W. W. Norton, 1977), p. 42.

16. A. C. Chakrabarti et al., "Synthesis of RNA by a Polymerase Protein Encapsulated Within Phospholipid Vesicles," *J Mol Evol* 39:555–59 (1994).

17. Deamer, pp. 239–61.

18. G. Sessa and G. Weissmann, "Phospholipid Spherules (Liposomes) as a Model for Biological Membranes," *J Lipid Res* 9:310–18 (1968).

19. Perlman, p. A2.

20. Sessa and Weissmann, pp. 310–18.

21. Sawyer, p. A1.

22. Deamer, pp. 239–61.

23. Dworkin et al., pp. 815–19.

24. Primo Levi, "Carbon," in *The Periodic Table* (New York: Schocken, 1984), p. 217.

25. Perlman, p. A2.

26. http://www.amacad.org/publications/bull6_ood.htm.

JANUARY 24, 2001

1. Scott Lindlaw, "Cowboy Boot Chic Has Arrived," AP, January 19, 2001, Washington.

2. "President: I Ask You to Be Citizens," *New York Times*, January 22, 2001, p. 14.

3. AP, "Survivors of Quake Move On to Cleanup as Aftershocks Fade," *St. Louis Post-Dispatch*, January 21, 2001, p. 10. Sylvia Moreno, "Area Salvadorans Learning of Lives, Homes Lost: News of Earthquake's Devastation in Homeland Trickling In," *Washington Post*, January 18, 2001, p. B1.

4. Elizabeth Tracey, "Dengue, Diarrhea Likely in El Salvador," Reuters, January 17, 2001.

5. Saul Hudson, "El Salvador Fights Disease, Quake Trauma," Reuters, January 18, 2001, Santa Tecla, El Salvador.

6. http://www.paho.org/English/SHA/be_v21n4-dengue.htm.

7. R. E. Isturiz et al., "Dengue and Dengue Hemorrhagic Fever in Latin America," *Infect Dis Clin North Am* 14(1):121–40, ix (March 2000).

8. Will Weissert, "Latin America Battles Lethal Fever," *Washington Post*, December 15, 2000, p. A54.

9. Clyde Haberman, "What Would Atheists Do? Join the Choir?" *New York Times*, March 7, 2000, p. B1.

10. Voltaire, *Candide* (Paris: Nouveau Classiques Larousse, 1970), p. 48.

11. Gary MacEoin, "U.S. Troops to El Salvador," *National Catholic Reporter* (Kansas City), October 13, 2000.

12. J. A. Rawlings et al., "Dengue Surveillance in Texas, 1995," *Am J Trop Med Hyg* 59(1):95–99 (July 1998).

13. J. Wilson, "Dengue Fever Moves North," *Popular Mechanics* 178:26 (January 2001).

14. Donald G. Mcneil Jr., "Hovering Where Rich and Poor Meet: The Mosquito," *New York Times*, September 3, 2000, p. 4.1.

15. S. Thein et al., "Changes in Levels of Anti-dengue Virus IgG Subclasses in Patients with Disease of Varying Severity," *J Med Virol* 40:102–6 (1993).

16. R. S. Rodriguez-Tan and M. R. Weir, "Dengue: A Review," *Tex Med* 94(10):53–59 (October 1998).

17. P. Avirutnan et al., "Dengue Virus Infection of Human Endothelial Cells Leads to Chemokine Production, Complement Activation, and Apoptosis," *J Immunol* 161:6338–46 (1998).

18. C. Kapp, "WHO Wins Reprieve for DDT Against Malaria," *Lancet* 356: 2076 (December 2000).

19. A. Attaran and R. Maharaj, "Doctoring Malaria, Badly: The Global Campaign to Ban DDT; DDT and Mosquitoes," *Brit Med Journal* 321:1403–5 (December 2000).

20. http://www.nobel.se/medicine/laureates/1948.

21. Michael Satchell and Don L. Boroughs, "Rocks and Hard Places: DDT: Dangerous Scourge or Last Resort?" *U.S. News & World Report* 129:64–65 (December 11, 2000).

22. R. Carson, *Silent Spring* (Cambridge, MA: Riverside Press, 1962), pp. 7–8.

23. Joseph L. Bast, Peter J. Hill, and Richard C. Rue, *Eco-Sanity: A Common-Sense Guide to Environmentalism* (Madison, WI: Madison Books, 1994), p. 100.

24. "Put Human Health First," *Detroit News*, December 20, 2000, p. 16.

25. M. S. Wolff and A. Weston, "Breast Cancer Risk and Environmental Exposure," *Envir Health Perspect* 105, suppl. 4:891–96 (June 1997).

26. N. E. Davidson and D. J. Yager, "Pesticides and Breast Cancer: Fact or Fad?" *J Natl Cancer Inst* 89:1743–44 (2000).

27. H. M. Blanck et al., "Age at Menarche and Tanner Stage in Girls Exposed in Utero and Postnatally to Polybrominated Biphenyl," *Epidemiology* 11:641–47 (2000).

28. Attran and Maharaj, p. 1404.

29. M. S. Wolff et al., "Organochlorine (OC) Exposures and Breast Cancer Risk in New York City Women," *Environ Res* 84(2):151–6 (October 2000).

30. R. Carson, *Always, Rachel: The Letters of Rachel Carson and Dorothy Freeman, 1952–1964: The Story of a Remarkable Friendship* (Boston: Beacon Press, 1995).

JANUARY 17, 2001

1. L. Thomas, *The Youngest Science* (New York: Viking, 1983), p. 112.

2. Marcia G. Synnott, *The Half-Opened Door: Discrimination and Admissions at Harvard, Yale, and Princeton, 1900–1970* (Westport, CT: Greenwood Press, 1979), p. viii.

3. G. Weissmann, *Democracy and DNA* (New York: Hill & Wang, 1995), p. 219.

4. Nitza Rosovsky, *The Jewish Experience at Harvard and Radcliffe: An Introduction to an Exhibition Presented by the Harvard Semitic Museum on the Occasion of Harvard's 350th Anniversary* (Cambridge, MA: The Semitic Museum, distributed by Harvard Univ. Press, 1986), p. 6.

5. Samuel E. Morison, *Three Centuries of Harvard 1636–1936* (Cambridge, MA: Harvard Univ. Press, 1937), p. 147.

6. Synnott, p. 18.

7. Morison, p. 147.

8. Rosovsky, p. 15.

9. Synnott, p. 21.

10. Ibid., p. 156.

11. G. Weissmann, personal recollection.

12. Henry K. Beecher and Mark D. Altschule, in *Medicine at Harvard: The First 300 Years* (Dartmouth, NH: New England Univ. Press, 1977), p. 479.

13. B. Sicherman, *Alice Hamilton: A Life in Letters* (Boston: Harvard Univ. Press, 1984), p. 234.

14. Joseph C. Aub and Ruth K. Hapgood, *Pioneer in Modern Medicine: David Lynn Edsall of Harvard* (Boston: Harvard Medical Alumni Association, 1970), p. 251.

15. Ibid, p. 251.

16. McG Harvey, *The Interurban Clinical Club 1905–1994* (Freehold, NJ: Interurban Clinical Club, 1995), appendix 1. Also Weissmann, p. 224.

17. Beecher and Altschule, p. 484.

18. Thomas, p. 260.

JANUARY 10, 2001

1. Richard Norton-Taylor, "Italy Blames Army Deaths on U.S. Shells," *Guardian* (Manchester), January 4, 2001, p. 12.

2. Massimo Giannini, "Per le morti fra i reduci chiedo conto alla Nato," *La Repubblica*, January 3, 2001, p. 1. http://www.repubblica.it/quotidiano/repubblica/20010103/interni/05amto.html.

3. Marlise Simons, "Radiation from Balkan Bombing Alarms Europe," *New York Times*, January 7, 2000, p. 3.

4. Norton-Taylor, p. 12.

5. Simons, p. 3.

6. D. P. Sandler, "Epidemiology of Acute Myelogenous Leukemia," *Seminr Oncol* 14:359–64 (1987).

7. Patricia Reaney, "Scientists Doubt Uranium Weapons Cancer Link," Reuters, January 9, 2001.

8. Norton-Taylor, p. 12.

9. Reuters, "Dutch Probe If Deaths Related to 'Balkan Syndrome,'" January 4, 2001. Also, Simons, p. 3.

10. Adrian Croft, "Belgium Wants EU to Discuss 'Balkans Syndrome,'" Reuters, December 29, 2000. Also Marlise Simons, "Europe to Investigate Uranium-Tipped Arms," *New York Times*, September 10, 2000.

11. http://www.repubblica.it/online/mondo/uraniodue/soldato/soldato.html.

12. http://www.guardianunlimited.co.uk/Archive/Article/0,4273,4113548,00.html.

13. Simons, *Europe*; Croft; Simons, *Radiation*, p. 3.

14. Simons, *Radiation*, p. 3.

15. W. Arkin, "A Politically Depleted Weapons," *Bull Atomic Scientists* 55(6): 729 (November/December 1999). http://www.thebulletin.org/issues/1999/nd99/nd99arkin.html.

16. Norton-Taylor, p. 12.

17. M. A. McDiarmid et al., "Health Effects of Depleted Uranium on Exposed Gulf War Veterans," *Environ Res* 82:168–80 (2000).

18. D. S. Rosenthal and H. J. Eyre, "Hodgkin's Disease and Non-Hodgkin's Lymphoma," in *American Cancer Society Textbook of Clinical Oncology* (Washington: American Cancer Society, 1995), pp. 456–57.

19. Arkin, p. 729.

20. T. C. Pellmar et al., "Electrophysiological Changes in Hippocampal Slices Isolated from Rats Embedded with Depleted Uranium Fragments," *Neurotoxicology* 20:785–92 (1999).

21. K. Birchard et al., "The Carcinogenic Effect of Localized Fission Fragment Irradiation of Rat Lung," *Int J Radiat Biol Relat Stud Phys Chem Med* 37:249–66 (1980).

22. A. C. Miller et al., "Transformation of Human Osteoblast Cells to the Tumorigenic Phenotype by Depleted Uranium-Uranyl Chloride," *Environ Health Perspect* 106:465–71 (1998). Also, A. C. Miller et al., "Suppression of Depleted Uranium-Induced Neoplastic Transformation of Human Cells by the Phenyl Fatty Acid, Phenyl Acetate: Chemoprevention by Targeting the p21RAS Protein Pathway," *Radiat Res* 155:163–70 (2001).

23. "Europe Gets Jittery about Old Uranium: U.S. Denies Cancer Link to High-Tech Weapons," *Houston Chronicle*, January 5, 2001, p. 14.

24. http://www.repubblica.it/quotidiano/repubblica/20010106/bari/03bula. html. http://www.repubblica.it/quotidiano/repubblica/20010107/esteri/ 11nato.html. "We Need an Inquiry into Depleted Uranium," leader, *Guardian* (Manchester), January 8, 2001, http://www.guardian unlimited. co.uk/Archive/Article/0,4273,4113742,00.html.

25. Arkin, p. 729.

26. B. N. Doebbeling et al., "Is There a Persian Gulf War Syndrome? Evidence from a Large Population-Based Survey of Veterans and Nondeployed Controls," *Am J Med* 108:695–704 (2000).

27. J. D. Knoke et al., "Factor Analysis of Self-Reported Symptoms: Does It Identify a Gulf War Syndrome?" *Am J Epidemiol* 152:379–88 (2000).

28. G. Weissmann, review of "The Harmony of Illusions," by Allan Young, *London Review of Books* 18:9–11, (1996). G. Weissmann, review of "From Paralysis to Fatigue: A History of Psychosomatic Illness in the Modern Era," by Edward Shorter, *New Republic*, April 6, 1992, pp. 38–41.

29. K. Douglas, "Aristocrats," in *Selected Poems* (London: Faber, 1964), p. 56.

DECEMBER 22, 2000

1. S. A. Austad, *Why We Age: What Science Is Discovering About the Body's Journey Through Life* (New York: John Wiley & Sons, 1997).

2. T. Kirkwood, *The Time of Our Lives: The Science of Human Aging* (Oxford: Oxford Univ. Press, 1999).

3. L. Hayflick, "Why Biogerontologists Should Not Write Popular Books on Aging," *Gerontologist* 38(4):508 (August 1998).

4. Lewis Thomas, "On Various Words," in *The Lives of a Cell* (New York: Viking Press, 1974), p. 128.

5. L. Thomas, "The Odds on Normal Aging," in *This Fragile Species* (New York: Macmillan, 1992), pp. 65–78.

6. Ibid.

DECEMBER 12, 2000

1. A. Bierce, *The Devil's Dictionary* (New York: A & C Boni, 1913).

2. Dow Jones, "Celera Genomics Submits Human Genome Manuscript," December 6, 2000, Rockville, MD. scipak-l@aaas.org: SCIPAK-L: *Science's Plan for Celera's Genomic Data*, December 6, 2000.

3. Ibid.

4. Gina Kolata, "Celera to Charge Other Companies to Use Its Genome Data," *New York Times*, December 8, 2000.

5. Ibid.

6. http://www.weizmann-usa.org.

7. S. Broder and J. C. Venter, "Sequencing the Entire Genomes of Free-Living Organisms: The Foundation of Pharmacology in the New Millennium," *Annu Rev Pharmacol Toxicol* 40:97–132 (2000).

8. http://www.ncbi.nlm.nih.gov/genome/guide/.

9. Terence Chea, "Celera Offers Gene-Diversity Database," *Washington Post*, September 14, 2000, p. E9.

10. Broder and Venter, pp. 97–132.

11. http://www.ncbi.nlm.nih.gov/genome/guide/

12. *Columbia Encyclopedia*, online at http://www.bartleby.com/65/ge/gene.html.

13. O. Lowy, unpublished lectures to a decade of NYU medical students in the 1950s.

14. R. C. Lewontin, S. Stone, and L. J. Kamin, *Not in Our Genes* (New York: Pantheon, 1984). Also, L. L. Cavalli-Sforza, *Genes, Peoples, and Languages*, trans. Mark Seielstad (New York: North Point Press/Farrar, Straus and Giroux, 2000).

15. Cavalli-Sforza, p. 8.

16. http://www.kff.com/frameprize.htm.

17. M. F. Hammer et al., "Jewish and Middle Eastern Non-Jewish Populations Share a Common Pool of Y-Chromosome Biallelic Haplotypes," *Proc Natl Acad Sci USA* 97(12):6769–74 (June 2000).

18. M. A. Jobling and C. Tyler-Smith, "New Uses for New Haplotypes the Human Y Chromosome: Disease and Selection," *Trends Genet* 16(8):356–62 (August 2000).

19. Ibid.

20. Hammer et al., pp. 6769–74.

21. Ibid.

22. W. H. Auden, "Fleet Visit," in *The Shield of Achilles* (London: Faber & Faber, 1955), p. 38.

NOVEMBER 27, 2000

1. Suzanne Daley, "Such a Deal: A Million Tons of Animal Parts, To Go," *New York Times*, November 26, 2000.

2. http://www.lemonde.fr/doss/0,2324,2460-1-QUO,00.html.

3. N. J. Andrews et al., "Incidence of Variant Creutzfeldt-Jakob Disease in the UK," *Lancet* 356(9228):481–82 (August 2000).

4. C. Oppenheim et al., "MRI and the Second French Case of vCJD," *Lancet* 356(9225):253–4 (July 2000).

5. Suzanne Daley, "Living with Mad Cows: The French and Fear Itself," *New York Times*, November 19, 2000.

6. "France Bans a Prized Delicacy to Reduce 'Mad Cow' Risk," *International Herald Tribune*, November 11, 2000, http://www.iht.com/articles/1077.htm.

7. Daley, "Mad Cows."

8. Tom Haines, "Some in Paris Panicky after Fears over Mad Cow Grow Industry Struggles to Create Confidence after Beef Sales Fall," *Boston Globe*, November 17, 2000. Also, Daley, "Mad Cows."

9. G. Lefebvre, *The French Revolution*, vol. 1, trans. E. M. Evanson (New York: Columbia Univ. Press, 1962), p. 128. Also, http://www.lemonde.fr/doss/0,2324,2460-1-QUO,00.html.

10. Ibid.

11. Daley, "Mad Cows."

12. J. A. Mastrianni and R. P. Roos, "The Prion Diseases," *Semin Neurol* 20(3):337–52 (2000).

13. Daley, "Mad Cows."

14. Peter Ford, "Feed Now Fuels French Industry: Amid a 'Mad Cow' Scare, Factories Find a Novel Way to Dispose of Suspect Animal Byproducts," *Christian Science Monitor*, November 22, 2000, p. 6.

15. http://www.fnclcc.fr/-dic/dico/p.htm#149.

16. Carol J. Williams, " 'Mad Cow' Disease Detected in Germany: Discovery of 2 Cases Spurs Calls for Mandatory Testing and Ban on Animal Parts in

Feed," *Los Angeles Times*, November 25, 2000. Also, http://www.independ-ent.co. uk/argument/Commentators/2000-11/lichfield271100.shtml.

17. http://www.independent.co.uk/argument/Commentators/2000-11/lichfield 271100.shtml.

18. *International Herald Tribune*, "France Bans Delicacy."

19. http://www.lemonde.fr/doss/0,2324,2460-1-QUO,00.html.

20. http://www.nobel.se/medicine/laureates/1997/prusiner-autobio.html.

21. Ibid.

22. S. B. Prusiner, "Novel Proteinaceous Infectious Particles Cause Scrapie," *Science* 216(4542):136–44 (April 1982).

23. Ibid.

24. H. Bueler et al., "Mice Devoid of PrP Are Resistant to Scrapie," *Cell* 73(7):1339–47 (July 1993).

25. S. Mouillet-Richard et al., "Signal Transduction Through Prion Protein," *Science* 289(5486):1925–28 (September 2000).

26. C. Weissmann and A. Aguzzi, "Bovine Spongiform Encephalopathy and Early Onset Variant Creutzfeldt-Jakob Disease," *Curr Opin Neurobiol* 7(5):695–700 (October 1997).

27. "Fifty-third Calypso," *Cat's Cradle* by K. Vonnegut (New York: Vintage, 1998).

28. L. Thomas, "Scabies and Scrapie," in *The Youngest Science* (New York: Viking Press, 1983) p. 233.

NOVEMBER 13, 2000

1. Sandra Blakeslee, "Pesticide Found to Produce Parkinson's Symptoms in Rats," *New York Times*, November 5, 2000.

2. Ibid.

3. J. T. Greenamyre et al., "Mitochondrial Dysfunction in Parkinson's Disease," *Biochem Soc Symp* 66:85–97 (1999).

4. G. C. Davis et al., "Chronic Parkinsonism Secondary to Intravenous Injection of Meperidine Analogues," *Psych Res* 1(3):249–54 (December 1979).

5. J. W. Langston et al., "Chronic Parkinsonism in Humans Due to a Product of Meperidine-Analog Synthesis," *Science* 219(4587):979–80 (February 1983).

6. Review in A. H. Schapira, "Evidence for Mitochondrial Dysfunction in Parkinson's Disease: A Critical Appraisal," *Mov Disord* 9(2):125–38 (March 1994).

7. Review in J. M. Gorell et al., "The Risk of Parkinson's Disease with Exposure to Pesticides, Farming, Well Water, and Rural Living," *Neurology* 50(5):1346–50 (May 1998).

8. Historical review in S. Takei, "Rotenone, an Active Constituent of Derris Root (*Derriselliptica Benth.*)," *Biochem Zeit* 157:1–15 (1925).

9. H. Fukami and M. Nakajima, "Rotenone and Rotenoids in Naturally Occurring Insecticides," ed. M. Jacobson and D. G. Crosby (New York: Dekker, 1971), pp. 71–79.

10. Nianbai Fang and John E. Casida, "Anticancer Action of Cube Insecticide: Correlation for Rotenoid Constituents between Inhibition of NADH: Ubiquinone Oxidoreductase and Induced Ornithinedecarboxylase Activities," *PNAS* 95(7):3380–84 (March 1998).

11. Greenamyre et al., pp. 85–97.

12. B. Ehrenreich, "Tell It to the Swordfish," *Progressive* (Madison) 64:15–16 (September 2000).

13. B. Ehrenreich, "Vote for Nader," *Nation*, August 21, 2000, p. 27.

14. P. Brendon, *The Dark Valley: A Panorama of the 1930s* (London: Jonathan Cape, 2000), p. 231.

15. James Dao, "The Green Party: 10,000 Turn Out to Hear Nader Urge 'Shift in Power,' " *New York Times*, November 6, 2000.

16. http://www.webcom.com/~paf/ereignis.html.

17. M. Heidegger, *Being and Time*, trans. J. MacQuarrie and E. Robinson (New York: HarperCollins, 1962). Also, http://thuban.ac.hmc.edu/~tbeckman/personal/Heidart.html.

18. G. P. Heidegger, *Phenomenology and the Essence of Technology*, http://www.abdn.ac.uk/cpts/gorner.htm.

19. M. E. Zimmerman, *Contesting Earth's Future: Radical Ecology and Postmodernity* (Berkeley, CA: Univ. of California Press, 1994).

20. Hans Sluga, *Heidegger's Crisis: Philosophy and Politics in Nazi Germany* (Cambridge, MA: Harvard Univ. Press, 1993).

21. L. Thomas, *The Youngest Science* (New York: Viking, 1983), p. 87.

22. C. Park, in *The Environment: Principles and Applications* (London: Routledge, 1997), http://geogmain.lancs.ac.uk/park/creation/ecology.htm.

23. Ehrenreich, "Vote for Nader," p. 27.

24. "Unsafe at Any Speed," *New Republic* 223:11 (November 6, 2000).

25. April Witt, "Florida Green Party Is 'on the Map'; Unrepentant Nader Voters Bask in Spotlight over Pivotal Role in Ballot Deadlock," *Washington Post*, November 13, 2000.

26. V. Klemperer, *I Will Bear Witness: A Diary of the Nazi Years 1933–1941* (New York: Random House, 1998), p. 80.

OCTOBER 30, 2000

1. AP, "Uganda's Ebola Death Toll Rises to 73," October 30, 2000.

2. C. J. Peters and A. S. Khan, "Filovirus Diseases," *Curr Top Microbiol Immunol* 235:85–95 (1999).

3. http://www.trinstitute.org/ojpcr/3_2westbrook.htm.

4. www.iht.com:80/IHT/TODAY/THU/IN/ebola.2.html.

5. Reuters, "Uganda's Ebola Death Toll Climbs to 68, WHO Says ET," October 27, 2000, Geneva.

6. Chris Tomlinson, "Ebola Traced to Woman," AP, October 19, 2000, Kabende Opong. Also, Simon Robinson, "A Trip Inside an African Hot Zone," *Time*, October 30, 2000.

7. http://www.odci.gov/cia/publications/factbook/geos/ug.html.

8. C. F. Basler et al., "The Ebola Virus VP35 Protein Functions as a Type I IFN Antagonist," *Proc Natl Acad Sci USA* 97:12289–94 (2000).

9. Review in L. A. Streether, "Ebola Virus," *Br J Biomed Sci* 56(4):280–84 (1999).

10. Z. Y. Yang et al., "Identification of the Ebola Virus Glycoprotein as the Main Viral Determinant of Vascular Cell Cytotoxicity and Injury," *Nat Med* 6(8):886–89 (August 2000).

11. R. W. Ruigrok et al., "Structural Characterization and Membrane Binding Properties of the Matrix Protein VP40 of Ebola Virus," *J Mol Biol* 300(1):103–12 (June 2000).

12. O. W. Holmes, "The Two Armies," in *Holmes's Complete Poetical Works* (Boston: Houghton Mifflin, 1908), p. 60.

OCTOBER 24, 2000

1. PR Newswire, "DeCODE Genetics and Roche Announce Identification of Schizophrenia Gene Friday," October 20, 2000, Reykjavik.

2. G. D. Pearlson, "Neurobiology of Schizophrenia," *Ann Neurol* 48(4):556–66 (October 2000).

3. J. R. Gulcher and K. Stefansson, "The Icelandic Healthcare Database and Informed Consent," *N Eng J Med* 342:1827–30 (June 2000).

4. Ibid.

5. Colin Nickerson, "Perfect for Genetic Research: Some Icelanders Are Wary of Losing Privacy: The Human Factor," *Boston Globe*, January 2, 2000.

6. http://www.decode.com.

7. Ibid.

8. S. F. Grant et al., "PCR Detection of the C4A Null Allele in B8-C4AQo-C4B1-DR3," *J Immunol Methods* 244(1–2):41–47 (October 2000).

9. Recent evidence from P. R. Taylor et al., "A Hierarchical Role for Classical Pathway Complement Proteins in the Clearance of Apoptotic Cells in Vivo," *J Exp Med* 192(3):359–66 (August 1, 2000).

10. M. E. Kemp et al., "Deletion of C4A Genes in Patients with Systemic Lupus Erythematosus," *Arthritis Rheum* 30(9):1015–22 (September 1987).

11. Nickerson.

12. Ibid.

13. Gulcher and Stefansson, pp. 1827–30.

14. PR Newswire, "CNW Vertex and Glaxo Wellcome's New Anti-HIV Medicine, Agenerase(TM) (Amprenavir), Approved in the European Union," October 23, 2000, Cambridge, Mass.

15. W. Bradford, *Of Plymouth Plantation*, ed. S. E. Morrison (New York: Knopf, 1952), p. 422.

16. S. E. Morrison, *The European Discovery of America* (New York and Oxford: Oxford Univ. Press, 1971).

OCTOBER 17, 2000

1. http://www.nobel.se/announcement/2000/medicine.html.

2. L. Büchner, *Kraft und Stoff* (Jena, Germany: Fischer Verlag, 1855)

3. A. Carlsson et al., "Neurotransmitter Aberrations in Schizophrenia: New Perspectives and Therapeutic Implications," *Life Sci* 61(2):75–94 (1997).

4. P. Svenningsson et al., "Dopamine D(1) Receptor-Induced Gene Transcription Is Modulated by DARPP-32," *J Neurochem* 75(1):248–57 (July 2000).

5. A. Casadio et al., "A Transient, Neuron-wide Form of CREB-Mediated Long-Term Facilitation Can Be Stabilized at Specific Synapses by Local Protein Synthesis," *Cell* 99(2):221–37 (October 1999).

6. S. Freud, "Some Points in a Comparative Study of Organic and Hysterical Paralysis," in *Collected Papers*, vol. 1, ed. E. Jones (1893; reprint, London: Hogarth Press, 1924), p. 54.

7. E. R. Kandel, "Cell Biology and the Study of Behavior," in *The Biological Revolution*, ed. G. Weissmann (New York: Plenum Press, 1979), pp. 79–102.

8. E. R. Kandel, "A New Intellectual Framework for Psychiatry," *Am J Psychiatry* 155(4):457–69 (April 1998).

9. S. Freud, "Heredity and the Aetiology of Neurosis," in *Collected Papers*, vol. 1, ed. E. Jones (1896; reprint, London: Hogarth Press, 1924), p. 143.

10. http://sde.caltech.edu/ellison.

11. C. M. McCay, M. F. Crowell, and L. A. Maynard, "The Effect of Retarded Growth upon the Length of Life-Span and upon the Ultimate Body Size," *J Nutrition* 10:63–79 (1935).

12. Nicholas Wade, "Scientist at Work: Dr. Leonard Guarente: Searching for Genes to Slow the Hands of Biological Time," *New York Times*, September 26, 2000. Also, S. J. Lin, P.-A. Defossez, and L. Guarente, "Requirement of NAD and SIR2 for Life-Span Extension by Calorie Restriction in Saccharomyces Cerevisiae," *Science* 289(5487):2126–28 (September 2000).

13. C. K. Lee, R. Weindruch, and T. A. Prolla, "Gene-Expression Profile of the Ageing Brain in Mice," *Nat Genet* 25(3):294–97 (July 2000).

14. BW HealthWire, "Celera Completes Sequencing of 9.3 Billion Base Pairs of Mouse Genome: Genetic Diversity between Mouse Strains Provides Extensive Mouse SNP Database," October 12, 2000, Rockville, MD.

15. A. Huxley, *After Many a Summer Dies the Swan* (New York: Harpers, 1939), p. 82.

16. Ibid.

17. http://www.huntington.org/EventsCal.html.

18. A. L. Tennyson, "Tithonus," in *The Poetic and Dramatic Works of A. L. Tennyson*, vol. 3 (New York: Houghton Mifflin, 1929).

OCTOBER 3, 2000

1. Gina Kolata, "U.S. Approves Abortion Pill," *New York Times*, October 29, 2000. Also, Aaron Zitner, "FDA Approves Use of Abortion Pill," *Los Angeles Times*, October 29, 2000.

2. Review in M. D. Creinin, "Medical Abortion Regimens: Historical Context and Overview," *Am J Obstet Gynecol* 183(2 Suppl.):S3–9 (August 2000).

3. M. Bygdeman et al., "Contraceptive Use of Antiprogestin," *Eur J Contracept Reprod Health Care* 4(2):103–7 (June 1999).

4. G. Sessa and G. Weissmann, "Differential Effects of Etiocholanolone on Phospholipid/Cholesterol Structures Containing Either Testosterone or Estradiol," *Biochim Biophys Acta* 150(2):173–80 (March 1968).

5. E. E. Baulieu, "Cell Membrane: A Target for Steroid Hormones," *Mol Cell Endocrinol* 12(3):247–54 (December 1978).

6. Review in E. E. Baulieu, "Contragestion and Other Clinical Applications of RU-486: An Antiprogesterone at the Receptor," *Science* 245:(4924) 1351–57 (1989).

7. W. Herrmann et al., "The Effects of an Anti-progesterone Steroid in Women: Interruption of the Menstrual-Cycle and of Early Pregnancy," *Comptes rendus de l'academie des sciences serie III: Sciences de la Vie* 294:(18) 933–38 (1982).

8. E. E. Baulieu and S. J. Segal, eds, *The Antiprogestin Steroid RU486 and Human Fertility Control* (New York: Plenum, 1985).

9. R. Kurzrock and C. C. Lieb, "Biochemical Studies of Human Semen II: The Action of Human Semen on the Human Uterus," *Proc Soc Eper Biol & Med* 28:268–73 (1930).

10. S. Bergstrom and J. Sjovall, "The Isolation of Prostaglandin E from Sheep Prostate Glands," *Acta Chem Sand* 14:1701–6 (1960).

11. M. Hamberg and B. Samuelsson, "On the Mechanism of the Biosynthesis of Prostaglandins E-1 and F-1-alpha," *J Biol Chem* 242(22):5336–43 (November 1967).

12. Description in G. Weissmann, "Nobel Week 1982," in *The Woods Hole Cantata* (New York: Dodd Mead, 1985), pp. 193–222.

13. C. Bernard, *Introduction to the Study of Experimental Medicine*, trans. H. C. Greene (New York: Schuman, 1947), p. 123.

SEPTEMBER 18, 2000

1. Alicia Ault, "30 Ill after Race through a Jungle," *New York Times*, September 14, 2000.

2. http://www.cdc.gov/od/oc/media/pressrel/r2k0913.htm.

3. http://www.ecochallenge.com/borneo2000.

4. S. J. Antony, "Leptospirosis: An Emerging Pathogen in Travel Medicine: A Review of Its Clinical Manifestations and Management," *J Travel Med* 3(2):113–18 (June 1996).

5. "Leptospirosis and Unexplained Acute Febrile Illness among Athletes Participating in Triathlons—Illinois and Wisconsin, 1998," *Morb Mortal Wkly Rep* 47(32):673–76 (August 1998).

6. C. Stephan et al., "Leptospirosis Illnesses after a Staff Outing," *Deutsch Med Wochenschr* 125(19):623–27 (May 2000).

7. T. Thiruventhiran and S. Y. Tan, "The Patient Who Had a Picnic at a Waterfall and Presented with Haemoptysis and Renal Failure," *Nephrol Dial Transplant* 15(5):727–78 (May 2000).

8. http://www.ecochallenge.com/borneo2000/news and updates.

9. Chris Burrell, "CDC Combs Vineyard for Disease Clues, Fatal Illness Wet Weather May Be Linked," *Boston Globe*, September 10, 2000.

10. Chris Burrell and Ellen Barry, "Vineyarders Cautious over 'Rabbit Fever,' " *Boston Globe*, September 2, 2000.

11. J. F. Levine and A. Spielman, "Effect of Deer Reduction on Abundance of the Deer Tick (*Ixodes dammini*)," *Yale J Biol Med* 57(4):697–705 (July 1984).

12. A. Spielman, "The Emergence of Lyme Disease and Human Babesiosis in a Changing Environment," *Ann NY Acad Sci* 740:146–56; 57: 697–705 (1994).

13. Ford Madox Ford, *Memories and Impressions*, vol. 5 (London: Bodley Head, 1962), p. 190.

14. P. Kropotkin, *Mutual Aid: A Factor of Evolution* (1902; reprint, New York: New York Univ. Press, 1972), p. 43.

15. P. Kropotkin, *Memoirs of a Revolutionist* (Boston and New York: Houghton Mifflin, 1899), p. 498.

16. Kropotkin, *Mutual Aid*, p. 43.

SEPTEMBER 5, 2000

1. Luke Baker, "Pope Tells Scientists Cloning Morally Unacceptable," Reuters, August 29, 2000, Rome.

2. Don Latin, "Vatican Assails New Guidelines on Human Embryo Research: Ethicists Divided over Morality of Cell Studies," *San Francisco Chronicle* August 25, 2000.

3. http://www.vatican.va accessed August 28, 2000.

4. Baker.

5. http://www.corriere.it accessed August 29, 2000.

6. Nicholas Wade, "New Rules on Use of Human Embryos in Cell Research," *New York Times*, August 24, 2000.

7. A. Colman and A. Kind, "Therapeutic Cloning: Concepts and Practicalities," *Trends Biotechnol* 18(5):192–96 (May 2000). Also, T. Asahara, C. Kalka, and J. M. Isner, "Stem Cell Therapy and Gene Transfer for Regeneration," *Gene Ther* 7(6):451–57 (March 2000).

8. I. Wilmut, L. Young, and K. H. Campbell, "Embryonic and Somatic Cell Cloning," *Reprod Fertil Dev* 10(7–8):639–43 (1998).

9. Ibid.

10. Jacques Barzun, *From Dawn to Decadence:500 Years of Western Cultural Life, 1500 to Present* (New York: HarperCollins, 2000), p. 796.

11. Latin.

12. Ibid.

13. Wade.

14. Richard M. Titmuss, *The Gift Relationship: From Human Blood to Social Policy* (New York: New Press, 1997).

AUGUST 22, 2000

1. S. I. Grewal, "Transcriptional Silencing in Fission Yeast," *J Cell Physiol* 184(3):311–18 (September 2000).

2. http://www.ellison-med-fn.org.

3. L. J. Wang, A. Kuzmich, and A. Dogariu, "Gain-Assisted Superluminal Light Propagation," *Nature* 406:277–79 (2000).

4. Thomas H. Maugh II, "Scientists Form Brain Cells from Bone Marrow," *Los Angeles Times*, August 15, 2000.

5. D. Woodbury et al., "Adult Rat and Human Bone Marrow Stromal Cells Differentiate into Neurons," *J Neurosci Res* 61(4):364–70 (August 2000).

6. P. Phelps, D. J. Prockop, and D. J. McCarty, "Crystal-Induced Inflammation in Canine Joints: Evidence against Bradykinin as a Mediator of Inflammation (Hopkins Memorial Medal lecture)," *J Lab Clin Med* 68(3):433–47 (September 1968).

7. D. J. Prockop, "Pleasant Surprises en Route from the Biochemistry of Collagen to Attempts at Gene Therapy (Hopkins Memorial Medal lecture)," *Biochem Soc Trans* 27(2):15–31 (February 1999).

8. C. R. Bjornson et al., "Turning Brain into Blood: A Hematopoietic Fate Adopted by Adult Neural Stem Cells in Vivo," *Science* 283(5401):534–47 (January 1999).

9. Maugh II.

10. Woodbury et al., pp. 364–70.

11. Maugh II.

12. Woodbury et al., pp. 364–70.

13. P. J. Pauly, *Controlling Life: Jacques Loeb and the Engineering Ideal in Biology* (Oxford: Oxford Univ. Press, 1987), p. 100.

14. Ibid., p. 103.

AUGUST 8, 2000

1. M. Cordero-Erausquin et al., "Nicotinic Receptor Function: New Perspectives from Knockout Mice," *Trends in Pharmacological Sciences* 21:211–17 (2000).

2. C. Stough et al., "Smoking and Raven IQ," *Psychopharmacology* 116(3):382–84 (November 1994).

3. R. Peto et al., "Smoking, Smoking Cessation, and Lung Cancer in the UK," *BMJ* 321(7257):323–29 (August 2000).

4. F. Baker, S. R. Ainsworth, and J. T. Dye "Health Risks Associated with Cigar Smoking," *JAMA* 284:735–40 (2000).

5. F. M. Brennan, R. N. Maini, and M. Feldmann, "TNF Alpha: A Pivotal Role in Rheumatoid Arthritis?" *Br J Rheumatol* 31(5):293–98 (May 1992).

6. http://www.newscientist.com:80/news/news_225038.

7. A. M. Malfait et al., "The Nonpsychoactive Cannabis-Constituent Cannabidiol Is an Oral Anti-arthritic Therapeutic in Murine Collagen-Induced Arthritis," *Proc Natl Acad Sci USA* 97(17):9561–66 (August 2000).

8. A. J. Hampson et al., "Cannabidiol and (-)Delta9-Tetrahydrocannabinol Are Neuroprotective Antioxidants," *Proc Natl Acad Sci USA*, 95(14):8268–73 (July 1998).

9. "Cannabinoid Antioxidant Protects Brain Cells—without the High," http://www.nimh.nih.gov/events/prcann.htm.

JULY 27, 2000

1. A. J. Liebling, *Between Meals: An Appetite for Paris* (New York: Simon & Schuster, 1967), p. 10.

2. Ibid.

3. M. De Lorgeril, P. Salen, and J. L. Martin, "Mediterranean Diet, Traditional Risk Factors, and the Rate of Cardiovascular Complications after Myocardial Infarction: Final Report of the Lyon Diet Heart Study," *Circulation* 99(6):779–85 (February 1999). C. Bosetti et al., "Foodgroups and Risk of Squamous Cell Esophageal Cancer in Northern Italy," *Int J Cancer* 87(2):289–94 (July 2000).

4. F. O. Ayorinde, K. Garvin, and K. Saeed "Determination of the Fatty Acid Composition of Saponified Vegetable Oils Using Matrix-Assisted Laser Desorption/Ionization Time-of-Flight Mass Spectrometry," *Rapid Commun Mass Spectrom* 14(7):608–15 (2000).

5. D. Palli et al., "Diet, Metabolic Polymorphisms and DNA Adducts: The EPIC-Italy Cross-sectional Study," *Int J Cancer* 87(3):444–51 (August 1987).

6. De Lorgeril et al., pp. 779–85. Bosetti et al., pp. 289–94.

7. C. Bosetti et al., "Fraction of Prostate Cancer Incidence Attributed to Diet in Athens, Greece," *Eur J Cancer Prev* 9(2):119–26 (April 2000).

8. A. Routtenberg et al., "Enhanced Learning after Genetic Overexpression of a Brain Growth Protein," *Proc Natl Acad Sci USA* 97(13):7657–62 (June 2000).

9. D. J. Linden, K. Murakami, and A. Routtenberg, "A Newly Discovered Protein Kinase C Activator (Oleic Acid) Enhances Long-Term Potentiation in the Intact Hippocampus," *Brain Res* 379(2):358–63 (August 1986).

10. S. K. Manna, A. Mukhopadhyay, and B. B. Aggarwal, "Resveratrol Suppresses TNF-Induced Activation of Nuclear Transcription Factors NF-KappaB Activator Protein-1, and Apoptosis: Potential Role of Reactive Oxygen Intermediates and Lipid Peroxidation," *J Immunol* 164(12):6509–19 (June 2000).

11. M. Holmes-McNary and A. S. Baldwin Jr., "Chemopreventive Properties of Trans-resveratrol Are Associated with Inhibition of Activation of IkB Kinase," *Cancer Research* 60:3477–83 (July 2000).

12. J. J. Moreno, "Resveratrol Modulates Arachidonic Acid Release, Prostaglandin Synthesis, and 3T6 Fibroblast Growth," *J Pharmacol Exp Ther* 294(1):333–38 (July 2000).

13. B. N. Cronstein and G. Weissmann, "Targets for Antiinflammatory Drugs," *Annu Rev Pharmacol Toxicol* 35:449–62 (1995).

JULY 20, 2000

1. Anne Gearan, "Funds for Alzheimer's Provided," AP, July 16, 2000.

2. D. Schenk et al., "Immunization with Amyloid-Beta Attenuates Alzheimer-Disease-like Pathology in the PDAPP Mouse," *Nature* 400(6740):173–77 (July 1999).

3. Lauran Neergaard, "Alzheimer's Vaccine Safe So Far," AP, July 11, 2000.

4. J. Madeleine Nash, "The New Science of Alzheimer's," *Time* 156(3) July 17, 2000.

5. G. Kolata, "Separating Research from News," *New York Times*, July 18, 2000.

6. M. Pras et al., "Physical, Chemical, and Ultrastructural Studies of Water-Soluble Human Amyloid Fibrils: Comparative Analyses of Nine Amyloid Preparations," *J Exp Med* 130(4):777–96 (October 1969).

7. M. J. Saraiva, P. P. Costa, and D. S. Goodman, "Amyloid Fibril Protein in Familial Amyloidotic Polyneuropathy, Portuguese Type: Definition of Molecular Abnormality in Transthyretin (Prealbumin)," *J Clin Invest* 74(1):104–19 (July 1984).

8. L. Thomas, "Natural Science," in *The Lives of a Cell* (New York: Viking Press, 1974), p. 124.

9. http:www.usnews.com/usnews/nycu/health/hosptl/tophosp.htm.

10. H. G. Clark, *Outlines of a Plan for a Free City Hospital* (Boston: Ticknor & Fields, 1860).

11. L. Thomas, W. Pelz, and F. Ingelfinger, in *The Harvard Medical Unit at the Boston City Hospital*, eds. Max Finland and William B. Castle (Boston: Countway Library of Medicine HMS, 1983).

12. Lewis Thomas, *The Youngest Science* (New York: Viking, 1983), p. 36.

13. Thomas et al.

14. Ibid.

15. Ibid.

16. Thomas, p. 36.

17. Ibid, p. 27.

18. H. Zinsser, *Rats, Lice and History* (Boston: Little, Brown, 1935), p. 58.

19. H. Zinsser, *As I Remember Him: The Biography of R.S.* (Boston: Little, Brown, 1940), p. 441.

JULY 13, 2000

1. R. Swarns and L. Altman, "AIDS Forum in South Africa Opens Knotted in Disputes," *New York Times*, July 10, 2000.

2. P. Duesberg and D. Rasnick, "The AIDS Dilemma: Drug Diseases Blamed on a Passenger Virus," *Genetica* 104(2):85–132 (1998).

3. M. Fox, "AIDS Causes Falling Population in Africa," Reuters, July 10, 2000.

4. M. Fox, "Leading AIDS Researcher Attacks S. Africa's Mbeki," Reuters, July 11, 2000.

5. L. T. Kohn, J. M. Corrigan, and M. S. Donaldson, eds., *To Err Is Human* (Washington, DC: National Academy Press, 2000).

6. L. L Leape and D. M. Berwick, "Safe Health Care: Are We Up to It? *BMJ* 320(7237):725–76 (March 2000). Also, E. J. Thomas, E. J. Orav, and T. A. Brennan, "Hospital Ownership and Preventable Adverse Events," *J Gen Intern Med* 15(4):211–29 (April 2000).

7. T. A. Brennan, "The Institute of Medicine Report on Medical Errors: Could It Do Harm?" *N Engl J Med* 342(15):1123–35 (April 2000).

8. C. J. McDonald, M. Weiner, and S. L. Hui, "Deaths Due to Medical Errors Are Exaggerated in Institute of Medicine Report," *JAMA* 284(1):93–95 (July 2000).

JULY 4, 2000

1. P. O'Brian, *Blue at the Mizzen* (New York: W. W. Norton, 1999).

2. A. Foreman, "Still Doing Their Duty," *New York Times Book Review*, December 5, 1999, p. 30.

3. D. King, *Patrick O'Brian: A Life Revealed* (New York: Henry Holt, 2000).

4. Ibid.

5. P. O'Brian, *The Mauritius Command* (New York: W. W. Norton, 1977), p. 116.

6. P. O'Brian, *The Far Side of the World* (New York: W. W. Norton, 1994), pp. 145–46.

7. O'Brian, *Blue*.

8. W. F. Buckley Jr., "Eulogies: Patrick O'Brian," *Time*, January 17, 2000, p. 33.

9. D. Mamet, "The Humble Genre Novel, Sometimes Full of Genius: In Memoriam, Patrick O'Brian," *New York Times*, January 17, 2000, p. E1.

10. G. F. Will, "O'Brian Rules the Waves," *Washington Post*, January 13, 2000, p. A19.

JUNE 26, 2000

1. http://www.ornl.gov/hgmis/project/progress; http://www.nhgri.nih.gov.

2. http://www.celera.com.

3. http://www.sanger.ac.uk/HGP/draft2000/mainrelease.shtml.

4. Lewis Thomas, *Late Night Thoughts on Listening to Mahler's Ninth Symphony* (New York: Viking, 1983), p. 36.

5. "Capitol Hill Faces Possible Struggle with Genome Technology," CNN.com, June 26, 2000.

6. J. Watson and F. H. Crick, "Molecular Structure of Nucleic Acids: A Structure for Deoxyribosenucleic Acid," *Nature* 171:737–38 (April 1953).

7. N. Wade, "Genetic Code of Human Life Is Cracked by Scientists," *New York Times*, June 27, 2000.

8. Ibid.

9. J. Watson, "It's a Giant Resource That Will Change Mankind . . ." [sound file of interview], BBC News, http://news6.thdo.bbc.co.uk/hi/english/sci/tech/newsid%5F806000/806468.stm, accessed June 26, 2000.

10. R. W. Emerson, "The Adirondacs [sic]," in *Poems* (Cambridge, MA: Riverside Press, 1918), pp. 182–87.

JUNE 13, 2000

1. J. L. Nortier et al., "Urothelialcarcinoma Associated with the Use of a Chinese Herb (Aristolochia Fangchi)," *N Engl J Med* 342(23):1686–92 (June 2000).

2. D. A. Kessler, "Cancer and Herbs," *N Engl J Med* 342(23):1742–43 (June 8, 2000).

3. L. Greensfelder, "Herbal Product Linked to Cancer," *Science* 288(5473): 1946 (June 16, 2000).

4. P. E. Lipsky, X. L. Tao, "A Potential New Treatment for Rheumatoid Arthritis: Thunder God Vine," *Semin Arthritis Rheum* 26:713–23 (1997). D. W. Pyatt et al., "Hematotoxicity of the Chinese Herbal Medicine *Tripterygium wilfordii* Hook F of CD_{34}-positive Human Bone Marrow Cells," *Mol Pharmacol* 57:512 (2000).

5. E. Zola, "J' Accuse," *L'Aurore* (Paris), January 13, 1898, p. 1.

6. L. Neergaard, "Abortion Pill May Have Restrictions," AP, June 7, 2000.

7. J. G. Kahn et al., "The Efficacy of Medical Abortion: A Meta-Analysis," *Contraception* 61(1):29–40 (January 2000).

8. J. P. Guengant et al., "Mifepristone-Misoprostol Medical Abortion: Home Administration of Misoprostol in Guadeloupe," *Contraception* 60(3):167–72 (September 1999).

9. J. M. Drazen, "Asthma Therapy with Agents Preventing Leukotriene Synthesis or Action," *Proc Assoc Am Physicians* 111(6):547–59 (November/December 1999).

10. J. M. Drazen et al., "Pharmacogenetic Association between a5-LO Promoter Genotype and the Response to Anti-asthma Treatment," *Nature Genetics* 22(2):168–70 (June 1999).

11. J. M. Drazen, "Association between Genotypes in the 5-LO Pathway and Asthma Treatment Responses," Proc. 11, International Conference on Advances in Prostaglandin in and Leukotriene Research, Florence, Italy, June 4–8, 2000, p. 1.

MAY 23, 2000

1. L. O'Connor et al., "Apoptosis and Cell Division," *Curr Opin Cell Biol* 12:257–63 (2000).

2. V. A. Fadok et al., "A Receptor for Phosphatidylserine-specific Clearance of Apoptotic Cells," *Nature* 405:85–90 (2000).

3. Dwight Garner, "More Coffee? Wonder What She Meant by That?" review of *The Verificationist* by Donald Antrim, *New York Times Book Review*, February 20, 2000.

4. Edward Shorter, *From Paralysis to Fatigue: A History of Psychosomatic Illness in the Modern Era* (reprint, New York: Free Press, 1993).

5. L. Rangel et al., "The Course of Severe Chronic Fatigue Syndrome in Childhood," *Proc R Soc Med* 93:129–34 (2000).

6. L. Rangel et al., "Personality in Adolescents with Chronic Fatigue Syndrome," *Eur Child Adolesc Psychiatry* 9:39–45 (2000).

7. O. W. Holmes, *Medical Essays*, vol. 9, *Collected Works* (Boston: Riverside Press, 1892), p. 255.

8. Reuters, "Briton Cures Fatigue by Drilling Hole in Own Head," February 22, 2000, London.

9. W. B. Cannon, *The Wisdom of the Body* (Birmingham, AL: Leslie B. Adams Jr., 1932). W. B. Cannon, *The Way of an Investigator: A Scientist's Experiences in Medical Research* (New York: W. W. Norton, 1945).

10. A. Benison, C. Barger, and E. L. Wolfe, eds., *Walter B. Cannon: The Life and Times of a Young Scientist* (Cambridge, MA: Harvard Univ. Press, 1987), p. 405.

11. Ibid., p. 30.

12. W. B. Cannon, *Bodily Changes in Pain, Hunger, Fear and Rage: An Account of Recent Researches into the Function of Emotional Excitement* (New York: D. Appleton-Century, 1934).

13. W. B. Cannon, in *Homage à Charles Richet* by A. Pettit (Paris: Presse Institut, 1926), p. 9.

14. G. Weissmann, *Democracy and DNA* (New York: Hill & Wang, 1995), p. 140.

15. W. B. Cannon, "Social Homeostasis," *Science* 93:1–4 (August 1, 1941).

MAY 8, 2000

1. M. Cavazzana-Calvo et al., "Gene Therapy of Human Severe Combined Immunodeficiency (SCID)-X1 Disease," *Science* 288(5466):669–72 (April 2000).

2. R. Hirschhorn, "Adenosine Deaminase Deficiency: Molecular Basis and Recent Developments," *Clin Immunol Immunopathol* 76:S219–27 (September 1995).

3. J. B. Wyngaarden and E. W. Holmes Jr., "Molecular Nature of Enzyme Regulation in Purine Biosynthesis," *Ciba Found Symp* (48):43–64 (1977).

4. http://www.fda.gov.

5. Deborah Nelson, "Patient's Death in Gene Tests Not Reported: FDA Cites Many Other Lapses," *Washington Post*, May 3, 2000, p. A01. Also, Philip J. Hilts, "F.D.A. Says Researchers Failed to Report a Second Death Linked to Gene Therapy," *New York Times*, May 4, 2000, p. A20.

6. C. Bauters et al., "Site-specific Therapeutic Angiogenesis after Systemic Administration of Vascular Endothelial Growth Factor," *J Vasc Surg* 21(2):314–24 (February 1995).

7. Reuters Health, May 4, 2000.

8. G. Sessa and G. Weissmann, "Phospholipid Spherules (Liposomes) as a Model for Biological Membranes," *J Lipid Res* 9(3):310–18 (May 1968).

9. A. Domashenko, A. Gupta, and G. Cotsarelis, "Efficient Delivery of Transgenes to Human Hair Follicle Progenitor Cells Using Topical Lipoplex," *Nature Biotechnology* 18(4):420–23 (April 2000).

10. Ibid.

APRIL 3, 2000

1. AP, "Dietary Supplement Claims Debated," March 31, 2000, Gaithersburg, MD.

2. J. I. Boullata and A. M. Nace, "Safety Issues with Herbal Medicine," *Pharmacotherapy* 20(3):257–69 (March 2000).

3. A. Fugh-Berman, "Herb-Drug Interactions," *Lancet* 355(9198):134–38 (January 2000).

4. F. Ruschitzka et al., "Acute Heart Transplant Rejection Due to Saint John's Wort," *Lancet* 355(9203):548–49 (February 2000). S. C. Piscitelli et al., "Indinavir Concentrations and St. John's Wort," *Lancet* 355(9203):547–48 (February 2000).

5. O. W. Holmes, "Currents and Countercurrents in Medical Science" in *The Complete Works of Oliver Wendell Holmes*, vol. 10, Medical Essays (Boston: Riverside Press, 1892), p. 230.

6. J. Page and D. Henry, "Consumption of NSAIDs and the Development of Congestive Heart Failure in Elderly Patients," *Arch Int Med* 160(6):777–84 (2000).

7. G. Weissmann, "NSAIDs: Aspirin and Aspirin-like Drugs," in *Cecil's Textbook of Medicine*, ed. L. Goldman and J. C. Bennett (Philadelphia: W. B. Saunders, 2000), p. 114. Z. Wang and P. Brecher, "Salicylate Inhibition of

Extracellular Signal-regulated Kinases and Inducible Nitric Oxide Synthase," *Hypertension* 34(6):1259–64 (December 1999). B. N. Cronstein, M. C. Montesinos, and G. Weissmann, "Salicylates and Sulfasalazine, but Not Glucocorticoids, Inhibit Leukocyte Accumulation by an Adenosine-dependent Mechanism That Is Independent of Inhibition of Prostaglandin Synthesis and p105 of NF Kappa B," *Proc Natl Acad Sci USA* 96(11):6377–81 (May 1999).

8. Ibid.

9. Scott W. Rowlinson et al., "Spatial Requirements for 15-(R)-Hydroxy-5Z,8Z,11Z,13E-eicosatetraenoic Acid Synthesis within the Cyclooxygenase Active Site of Murine COX-2; Why Acetylated Cox-1 Does Not Synthesize 15-(R)-HETE," *J Biol Chem* 275(9):6586–91 (March 2000).

10. J. G. Filep et al., "Anti-inflammatory Actions of Lipoxin A(4) Stable Analogs Are Demonstrable in Human Whole Blood: Modulation of Leukocyte Interactions," *J Periodontal Res* 34(7):370–73 (October 1999).

Acknowledgments

MOST OF THE ESSAYS in this book are extensions of columns written for the innovative e-publication *Praxis Post*, the "webzine of medicine and culture" (http://praxis.md/post), edited by Ivan Oransky, MD. I am most grateful to Sarah Greene, president/CEO of Praxis Press, the parent web site of *Praxis Post*, and to Jill Neimark, contributing editor of *Praxis Post* for getting me involved in this valuable enterprise in the first place. But I am most indebted to Dr. Oransky, and to his two able colleagues, David Krasnow, senior editor, and Stacy Boyd, assistant editor, for shaping my rough-hewn reports into neater columns, which can support the rigors of e-publication. They preside over a Web site that unites not only the two cultures but also the literary and electronic sensibilities. Their work has taught me that the e-format has its own rules that clash with those of the printed word at many points other than length alone. Brought up in the age of the informal essay, and accustomed to a format that permits a cozy mix of error and foresight, I present the dated entries in their more or less original form without the smug redress of hindsight. As before, I would like to thank Cathy Norton and the staff of the Marine Biological Laboratory-Woods Hole Oceanographic Institution library for their inestimable help,

Acknowledgments

Mme. Arlette Gaillet for her valuable library on the Ile St. Louis, and Andrea Cody of the Biotechnology Study Center of the NYU School of Medicine, without whose aid these pieces would not have been reduced to print. Continuing thanks are also due to Gloria Loomis, my literary agent, and to Erika Goldman, my editor, who has been an intrepid supporter and critic of my work for the last few years. Finally, I extend my apologies to members of my clan who have been involuntarily exposed to early versions of these essays over the dinner table: Lisa, Andrew, Debra, Susan, and Benjamin.

Index

ABOUT THE AUTHOR

A RECIPIENT of a Guggenheim Fellowship and two Rockefeller residencies at Bellagio, Gerald Weissmann, MD, is professor of medicine and director of the Biotechnology Study Center at New York University School of Medicine. His essays and reviews have appeared in the *New Republic*, the *London Review of Books*, and the *New York Times Book Review*, and have been collected in six volumes, most recently *Darwin's Audubon*. Dr. Weissmann lives in New York City and Woods Hole, Massachusetts.